Animal Osmoregulation

Oxford Animal Biology Series

Titles

Energy for Animal Life
R. McNeill Alexander

Animal Eyes
M. F. Land, D-E. Nilsson

Animal Locomotion
Andrew A. Biewener

Animal Architecture
Mike Hansell

Animal Osmoregulation
Timothy J. Bradley

The Oxford Animal Biology Series publishes attractive supplementary textbooks in comparative animal biology for students and professional researchers in the biologiacal sciences, adopting a lively, integrated approach. The series has two distinguishing features: first, book topics address common themes that transcend taxonomy, and are illustrated with examples from throughout the animal kingdom; and second, chapter contents are chosen to match existing and proposed courses and syllabuses, carefully taking into account the depth of coverage required. Further reading sections, consisting mainly of review articles and books, guide the reader into the more detailed research literature. The Series is international in scope, both in terms of the species used as examples and in the references to scientific work.

Animal Osmoregulation

Timothy J. Bradley

OXFORD
UNIVERSITY PRESS

Great Clarendon Street, Oxford OX2 6DP

Oxford University Press is a department of the University of Oxford.
It furthers the University's objective of excellence in research, scholarship,
and education by publishing worldwide in

Oxford New York

Auckland Cape Town Dar es Salaam Hong Kong Karachi
Kuala Lumpur Madrid Melbourne Mexico City Nairobi
New Delhi Shanghai Taipei Toronto

With offices in

Argentina Austria Brazil Chile Czech Republic France Greece
Guatemala Hungary Italy Japan Poland Portugal Singapore
South Korea Switzerland Thailand Turkey Ukraine Vietnam

Oxford is a registered trade mark of Oxford University Press
in the UK and in certain other countries

Published in the United States
by Oxford University Press Inc., New York

First published 2009

British Library Cataloguing in Publication Data

Data available

Library of Congress Cataloging in Publication Data

Data available

Typeset by Newgen Imaging Systems (P) Ltd., Chennai, India
Printed in Great Britain
on acid-free paper by
CPI Antony Rowe, Chippenham, Wiltshire

ISBN 978–0–19–856995–4 (Hbk.) 978–0–19–856996–1 (Pbk.)

10 9 8 7 6 5 4 3 2 1

Preface

Water is fundamental to life and all metabolic reactions are influenced by the aqueous environment in which they occur. The maintenance of an appropriate aqueous environment for metabolic reactions is critical at the molecular, cellular, and organismal level. At the molecular level, the configuration and function of proteins is strongly influenced by the activity of water and the concentrations and nature of ions present. At the cellular level, osmoregulation assures the maintenance of a proper milieu in the cell cytoplasm as well as the regulation of cell volume. At the organismal level, osmoregulatory organs provide for the uptake and retention of water in dry environments, and the elimination of excess water under dilute external conditions. Each of these three levels requires specific and sometimes distinct regulatory processes.

Water balance is a fundamental aspect of cell volume regulation. As the active transport of water does not occur in biologic systems, the active transport of ions is central to the transfer and the regulation of water in biologic systems. The transport of ions and the maintenance of ion gradients is one of the oldest and most fundamental functions of cell membranes. Ion gradients across membranes also serve in energy storage and transduction. These functions are therefore not only inextricably linked to osmoregulation and water balance, but also they are fundamental in the production, storage, transduction, and use of energy in animals.

Contents

1 The properties of water

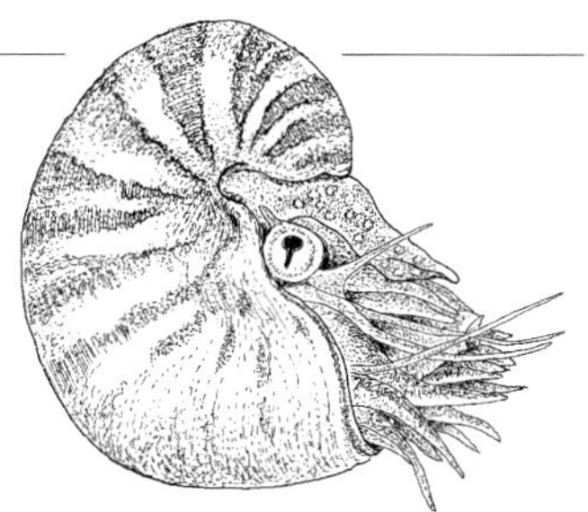

1.1 Introduction

A description of osmotic regulation in animals involves the mechanisms by which animal cells regulate the quantity and concentration of water in the body and in their cells. The need to regulate the volume of water is inherently clear. Every animal has a finite volume and the accumulation of more and more water eventually overwhelms the capacity of the tissues and organs to accommodate this increasing volume. Conversely, if water is lost to the environment in an unregulated manner, body volume will shrink until physiological functions, such as circulation, respiration, and locomotion, are adversely affected. Similarly, cells require an appropriate volume of cytoplasm to distribute ions and nutrients, facilitate cell motility, and retain the appropriate spacing of cellular organelles.

A second aspect of water is as important, perhaps more so, than simple volume. That aspect is the capacity of water to act as a solvent. The role of water as a solvent is vital for the proper configuration and function of proteins, sugars, lipids, and nucleic acids. Water is an essential aspect of every living organism. The discovery of water on Mars has excited those scientists who hoped to find evidence of life on Mars because water is absolutely essential to life as we know it. As we will see throughout this book, water alone is not sufficient for life. Living systems control and manipulate water in specific ways to accommodate vital activities. In this chapter, we will examine the properties of water and the means by which organisms can change those properties to facilitate a variety of vital activities.

1.2 The structure of pure water

Every elementary student is presented with the formula for water: H_2O. This formula indicates that every molecule of water consists of one atom of oxygen which is associated with two atoms of hydrogen. The oxygen atom sits in the middle of the molecule with the hydrogen atoms attached to it by covalent bonds on opposite sides of the oxygen. The covalent bonds are not precisely opposite to each other. That is, they are not 180° apart but are slightly off of a straight line such that the bonds form an angle ranging from 104° to 109°, depending on the state of the water (Fig. 1.1). As a result, the molecule has a "sidedness" with one side being most influenced by the presence of the hydrogen atoms, and the other side being dominated by the presence of an oxygen atom. The hydrogen and oxygen atoms do not share the electrons that form their covalent bonds equally. Instead, the electrons have a greater statistical likelihood of being near the oxygen nucleus than near the nuclei of the hydrogen atoms. As a result, the water molecule is a dipole, that is, one side of the molecule has a partial negative charge and the other side has a partial positive charge. This partial charge is usually

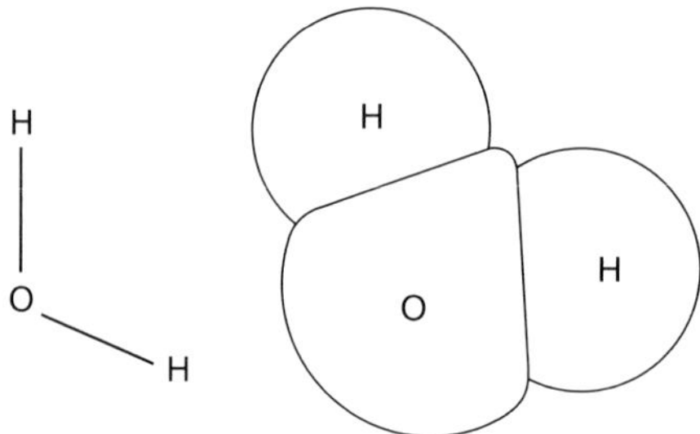

Fig. 1.1. Representations of the structure of a water molecule. Oxygen atoms are represented by an O, hydrogen atoms by a H. The left-hand figure is a "Ball and Stick" representation. Many chemists think the right-hand "Space-filling" model is a better representation of the actual structure of water.

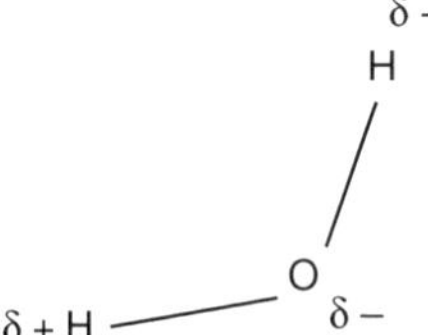

Fig. 1.2. A water molecule showing the distribution of partial charges across the molecule.

depicted with the Greek letter delta, as δ+ or δ− depending on the sign of the charge (Fig. 1.2). The charge is partial because the electron is not fully with one atom or the other. If the electron had moved over completely, it would leave the hydrogen nucleus with a charge of +1. As a result of the nonuniform charge distribution across the molecule, we refer to the water molecule as being polar.

When water is in its liquid or solid state, water molecules have the opportunity to come into close contact with each other. The polar nature of water influences the ways in which the water molecules interact. As like charges repel and unlike charges attract, adjacent water molecules have a preferred orientation with the negative side of one water molecule being more frequently associated with the positive side of the adjacent one. In liquid water, this can lead to a preferred orientation for a whole string of molecules such that each molecule affects the orientation of adjacent water molecules around it (Fig. 1.3). It is important to realize that in liquid water, the molecules are free to move about because of their kinetic energy. The orientation of water molecules with each other is therefore more accurately thought of as a favored or most probable orientation and not one that is fixed in time and space. In other words, the polar nature of water influences the statistical likelihood of the various orientations of water, with orientations in which the positive and negative poles that are adjacent being favored.

The associations of the positive and negative poles of water are fairly strong and these electrostatic attractions require energy to be broken. We refer to these associations as hydrogen bonding. At the temperatures at which water is liquid, kinetic energy serves to break the bonds fairly

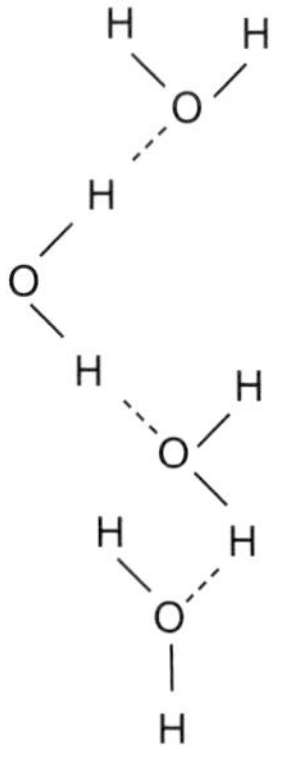

Fig. 1.3. Water molecules can be connected by hydrogen bonds between molecules that lead to an increase in the likelihood of the association of a partial negative charge on the oxygen atom being associated with the partial positive charge on an adjacent hydrogen atom. Covalent bonds are shown with a solid line, hydrogen bonds with a dotted line.

frequently, yet the energy contained in those bonds is substantial and has profound effects in biologic systems. Hydrogen bonding tends to glue water molecules together, increasing their adhesion to each other and influencing their configurations as illustrated in Figure 1.3.

1.3 Some unusual attributes of water

L. J. Henderson published a book in 1913 entitled *"The Fitness of the Environment"*. In his book, he argued that water has properties that make it unusually well suited as the solvent that supports life on Earth. Let us examine some of these properties.

1.3.1 Surface tension

Water has unusually strong surface tension. This is a phenomenon we see in the manner in which water adheres to and creeps up the side of a glass container, in the spherical shape of raindrops, and in the beading of water on waxy surfaces. When water molecules enter a narrow space they pull other water molecules along with them. This is an enormously important feature in the wetting of materials with water, which reaches its extreme in the case of trees that "pull" water up to the highest leaves along narrow xylem tubes. This can occur even in trees that are many meters high. The strong surface tension of water is the result of the strong intermolecular attraction between water molecules, a phenomenon associated with the hydrogen bonds as discussed above.

1.3.2 Density

Within the range of normal environments on Earth, water can exist as a solid, liquid, or gas depending on the temperature and pressure. The average distance between water molecules in the liquid state determines the density of water. As water becomes warmer, the molecules increase their kinetic energy and this increase causes the molecules to bang against each other a bit harder, putting more space between the rambunctious molecules. As a result, over most of the range in which water is liquid, the density of water goes down as temperature goes up. This has profound implications for biologic systems and ecosystems. For example, in the ocean, cold water tends to sink through warmer water, leading to the massive, three-dimensional current flows that drive the transfer of heat around the globe and through their actions can influence the distribution of nutrients in the oceans.

As water cools, it becomes denser and denser. This trend does not continue all the way to 0°C, the temperature at which water freezes. Instead, as water gets very near to its freezing temperature, it actually decreases in density. Figure 1.4 is a graph illustrating the density of water versus temperature. It can be seen that the density of water is maximum at about 4°C. The reason for this is the fact that water in its solid phase has a lower density than in liquid phase. This occurs because the molecules enter into a crystalline array during freezing. The distance between the water molecules is greater in the solid state than in the liquid state. Stated somewhat differently, in the liquid state the water molecules are free to move around and maintain a relatively close association. The formation of ice crystals locks the molecules into a solid array in which molecules are stabilized in a crystalline array with an average spacing greater than in the liquid state.

As one lowers the temperature of liquid water (i.e., reduces the kinetic energy of the molecules), the water molecules begin to develop a crystalline state of association with their neighbors. At temperatures just above freezing that state exists briefly before being broken by kinetic forces. Nonetheless, the repeated occurrence of even this momentary crystalline condition causes the molecules to be more widely spaced, thereby lowering the density as the molecules approach the fully solid condition at 0°C.

This characteristic of water, namely, the solid state is less dense than the liquid state, has profound implications in biology. Consider a pond in a climate where summer temperatures are well above the freezing point of water and winter temperatures are well below. As can be seen in Fig. 1.5a, in the summer the air temperature is warmer on average than the water

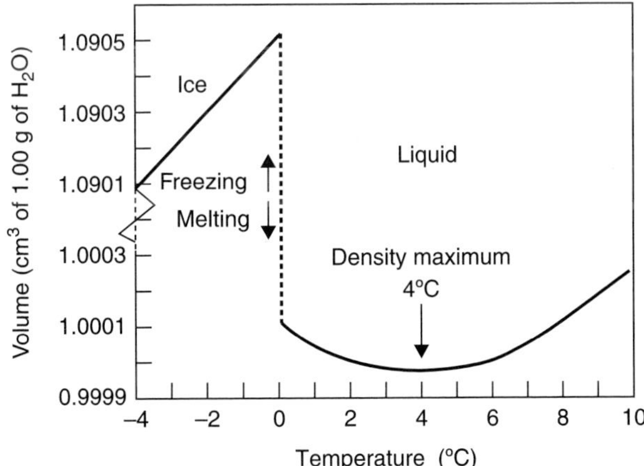

Fig. 1.4. The density of water as a function of temperature.

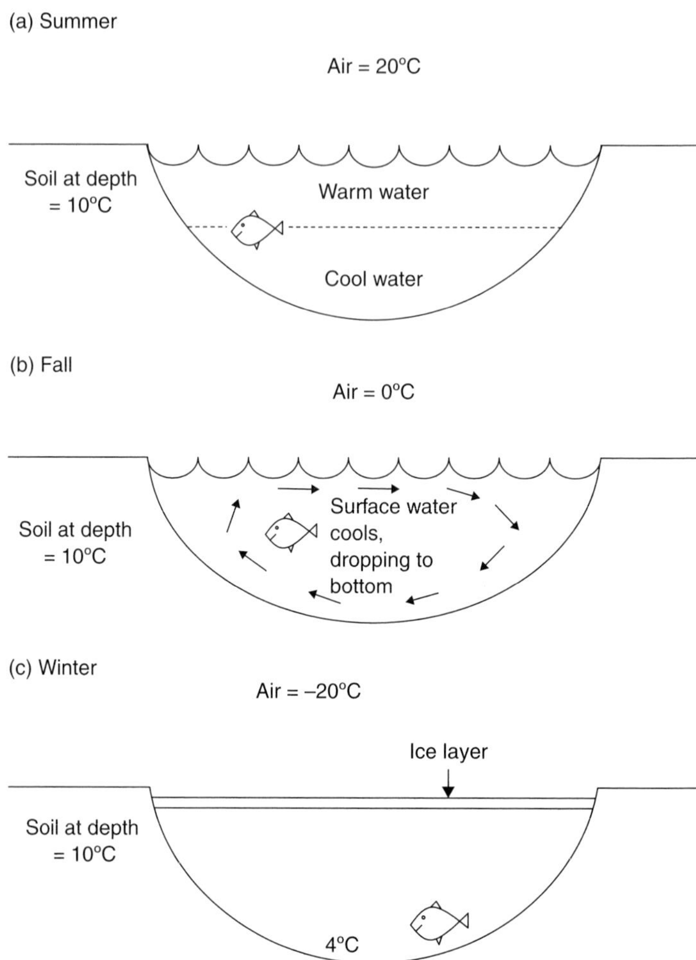

Fig. 1.5. Conditions of a pond in (a) summer, (b) fall, and (c) winter.

because of the thermal inertia of this relatively large body of water, because the ground is cooler than the air, and because of the effects of evaporation which cools the body of water. Figure 1.5b illustrates what happens in the fall when the air temperature drops. The water begins to cool at its upper surface, forming water more dense than that below. The pond is said to "turn over," that is, it tends to cool throughout as water cooled at the surface sinks all the way to the bottom, being now the densest water in the pond. To replace this sinking water, warmer water is brought up to surface where it in turn begins to cool. Figure 1.5c illustrates the conditions in winter. As fall proceeds into winter the water in the pond cools until it comes to a uniform 4°C. At this point, any water which cools further at the surface actually becomes less dense (note this effect in Fig. 1.4). This lighter water remains at the surface, cooling further and turning to ice. As this ice is less dense than liquid water, the ice floats and eventually forms a solid layer on the pond's surface. At this point, evaporative cooling is greatly reduced (ice does sublime but slowly). Any further cooling of the pond water can occur only by slow diffusion of heat from the pond water through the ice to the overlying cold air. The ground underlying the pond is not as cold as the overlying air because of the thermal inertia of the soil. In fact in most cold climates, the soil does not freeze down below a few centimeters. In Artic and Antarctic areas the ground freezes to a greater depth (still only a few meters) and does not fully thaw in the summer, a phenomenon known as permafrost.

To return to Fig. 1.5c, ice formed in the winter floats to the top and helps to seal the pond. As a result, as winter proceeds the ice may thicken if air temperatures remain cold, but under the ice the water is liquid. In fact, at the bottom of the pond, the temperature is a very precise 4°C, because this is the temperature at which water is most dense. The water at the bottom of the pond remains at 4°C if the winter is warm and the ice is thin, or if the winter is cold and the ice is thick. This allows animals, such as fish, frogs, turtles, and insects, to overcome winter near the pond bottom under conditions where the water temperature is stable and well above freezing. Of course, if very cold temperatures prevail, the pond can freeze all the way to the bottom, killing those organisms that are not freeze tolerant. In larger bodies of water, this is very unlikely to happen because the soil acts as a constant source of heat preventing the bottom water from reaching the freezing point.

Consider the alternative situation, if water in solid state were denser than water in liquid state. In such circumstances, ice would form in cold weather and sink to the bottom. The water would continue to be cooled by the overlying air and would circulate downward as it cooled. The warmest water would constantly be brought to the surface to be cooled. Under these conditions ponds would freeze much more rapidly and to a greater depth than they now do. Aquatic life in temperate climates would be much more difficult and larger freeze-sensitive organisms would probably be excluded altogether.

1.3.3 Vapor pressure

Water has a high heat of vaporization for a molecule of its size. The heat of vaporization is the amount of energy required to cause the molecule to change from its liquid phase to its gaseous phase. A high heat of vaporization means that a large amount of energy is required for this transition. Let us examine why water has a high heat of vaporization and why this is biologically important.

Imagine a beaker of water open at its surface to the atmosphere. The water has a certain temperature. This temperature reflects the average kinetic energy of the molecules in the beaker. If we remove energy from the water, its temperature will go down. If we add energy, the temperature will go up. Not all of the molecules in the beaker have equal kinetic energy. Some are moving fast, some are slow, but the average energy is a measurable value as reflected by the temperature. When a moving molecule approaches the air/water interface it may have sufficient kinetic energy to break free and enter the air as a single, gaseous molecule. In order to do that, it must have a trajectory that will carry it out into the air and sufficient energy to break the bonds of attraction with the adjoining molecules in the liquid. Because of the strong hydrogen bonds in liquid water, a substantial amount of energy is required to break free. As a result, at any given temperature water evaporates more slowly than other liquids of similar or even greater molecular weight. For example, the vapor pressure generated by water at 19°C is 16.5 mm of mercury, while the vapor pressure generated by ethanol at the same temperature is 40 mm of mercury. The higher heat of vaporization of water is equivalent to saying that water evaporates more slowly than ethanol and boils at a higher temperature.

The strong hydrogen bonds that exist in liquid water reduces the vapor pressure of water and raise the temperature at which the liquid boils. At boiling point, the vapor pressure of a liquid equals the pressure of the air around it (1 atm at sea level). Table 1.1 shows the boiling point of a number of substances in the size range of water. It can be seen that water has an unusually high boiling point.

Despite the strong hydrogen bonds holding the molecules together, water does evaporate from an open beaker, assuming that the air above it is not

Table 1.1. Molecular weight and boiling point of a number of substances in the size range of water

Compound	*Molecular weight*	*Temperature at which the vapor pressure equal 1 atm (°C)*
Carbon dioxide (CO_2)	44.01	−78.2
Ammonia (NH_3)	17.03	−33.6
Acetone (CH_3COCH_2)	58.08	56.5
Ethanol (CH_3CH_2OH)	46.07	78.4
Water (H_2O)	18.02	100

saturated with water vapor. Those molecules that do manage to escape into the air are unusual; however, they are the molecules with the greatest kinetic energy, as they managed to escape the force of the hydrogen bonds. In fact, all these molecules had kinetic energy above the average of the molecules in the beaker. As a result, when the high-energy molecules leave the water, this lowers the mean kinetic energy of the molecules in the beaker. As stated above, this means that evaporation of water from the surface tends to cool the water in the beaker. This is a phenomenon with which we all are familiar, evaporation cools the fluid or surface from which the evaporation occurs. This is true for every fluid. However, the large amount of energy required to separate each water molecule from its neighbor causes the energy that is removed per molecule to be unusually large compared to other fluids.

It can be seen that the water is an unusual molecule with interesting physical properties. Those properties have a profound effect on the living forms that have evolved on the largely aqueous surface of our planet. To this point in the chapter we have considered the properties of pure water. The interaction of water with solutes is also of critical importance in biologic systems. We will now turn our attention to those issues.

1.4 The colligative properties of water

Compounds which will dissolve in a liquid are referred to as solutes. To be considered a solute, the compound must interact with fluid and remain suspended in the absence of agitation. If you took a handful of soil and placed it in a beaker of water, it would, upon stirring, form a nice muddy mess. If you allowed this mixture to settle for a day and came back, you would find that you had formed a solution. Salts and other ionic compounds would dissolve in the water, dividing into their ionic forms. Organics such as sugar and proteins would also dissolve in the water. Colloids such as tiny clay particles might well also still be suspended and in solution. Larger particles of silt, sand, and mud would, however, have settled down. If we now decanted the overlying solution, we would have separated soluble compounds (i.e., solutes and colloids) from insoluble materials.

The presence of solutes profoundly affects the properties of liquid water. These properties, which are termed the colligative properties of water, include:

- The vapor pressure of the aqueous solution
- The freezing point of the aqueous solution
- The melting point of the aqueous solution
- The osmotic concentration of the aqueous solution

These properties change in concert because each is affected by the free energy of the water in solution. Let us examine this point in more detail.

1.4.1 Solute effects on vapor pressure

The effects of solutes on the free energy of water are most easily understood if we examine the effects of solutes on the vapor pressure of a solution. Consider the situation depicted in Figure 1.6, in which we have a closed container with an aqueous solution below and air above. Naturally over time, an equilibrium will occur in which the water vapor evaporating from the liquid will saturate the air above, leading to an equilibrium condition. This equilibrium is, by definition, the state in which the water has the same free energy in the liquid and vapor state. We can state this mathematically as

$$G_{\text{liquid}} = G_{\text{vapor}} \tag{1.1}$$

where G stands for the Gibbs free energy of the water molecules.

Solutes lower the vapor pressure of solutions as expressed by Raoult's law:

$$P_{\text{vapor}} = P_{\text{pure water}}(\text{X1}) \tag{1.2}$$

In this expression, the vapor pressure of an aqueous solution containing a solute (P_{vapor}) is equal to the vapor pressure of pure water ($P_{\text{pure water}}$) $\times$ the mole fraction of solvent (X1) present in the solution. The mole fraction of solvent (in this case water) is defined as

$$\text{X1} = \frac{\text{Moles of solvent}}{\text{Moles of solvent} + \text{Moles of solute}} \tag{1.3}$$

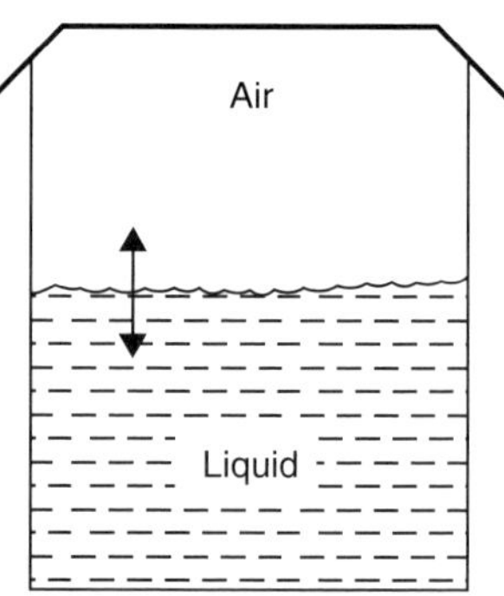

Fig. 1.6. A closed container in which an equilibrium has been established between the solvent in the liquid and the solvent in the gaseous state in the air.

The mole fraction is, therefore, a measure of the proportion of moles of that substance present in the solution. For example, if no solute was present in the solution, X1 would equal 1.0. If the solvent and solute were present in equimolar amounts, X1 would equal 0.5.

Raoult's law states that the vapor pressure of water over a solution declines as the percentage of water in that solution goes down, and this decline is directly proportional to the mole fraction of solvent. As the proportion of solvent molecules at the surface of the solution drops, the vapor pressure drops as well because fewer water molecules are available to evaporate from the surface.

The mole fraction of solvent is not the only relevant unit here; we can also refer to the mole fraction of solute. The mole faction of solvent (symbol X1) is a proportion which is always $\leqslant 1$. The mole fraction of solute (symbol X2) also is $\leqslant 1$. The mole fraction of solvent and mole fraction of solute when added together must equal 1.

$$X1 + X2 = 1 \tag{1.4}$$

Therefore,

$$X1 = 1 - X2 \tag{1.5}$$

If we substitute for X1 as defined in Eqn 1.2, we obtain

$$\frac{P_{\text{vapor}}}{P_{\text{pure water}}} = 1 - X2 \tag{1.6}$$

In other words, the reduction in vapor pressure is proportional to the mole faction of solute added. The more solute, the lower the vapor pressure.

We have made some good progress here. We have demonstrated that one of the very important properties of water varies inversely with the mole fraction of water present (or if you prefer directly with the mole fraction of solute present). However, mole fractions are units with which we have had little contact. To understand mole fractions a little more clearly, let us take a look at the mole fractional values of a solution with which we are familiar, namely seawater.

Seawater has about 1 mole of solutes per kilogram of water. It is a relatively concentrated solution by the standards of free-standing water on our planet. Only brine pools and salt lakes are more salty. Seawater is either equally salty or more salty than the body fluids of all animals. Seawater is, therefore, a useful example for us to use a relatively concentrated solution.

Given that seawater has 1 mole of solutes per kilogram of water, how many moles of water are present in a kilogram? Well, water has a molecular weight of 18 g so the number of moles of water in a kilogram is

$$\frac{1000\ \text{g}}{18\ \text{g/mole}} = 55.56 \tag{1.7}$$

Therefore, there are 55.56 moles of water in a kilogram of seawater and 1 mole of solute.

The mole fraction of solute in seawater is, therefore,

$$\frac{1}{55.55} = 0.018 \tag{1.8}$$

and it follows that the mole fraction of water in seawater is 0.982.

Given that we consider seawater to be a relatively concentrated solution from a biologic perspective (e.g., it is too concentrated for us to drink and remain in water balance), we can see that a relatively small reduction in the mole fraction of water can have a profound effect on the biologic properties of water.

As discussed above, the colligative properties of water vary in concert as a function of the mole fraction of solute present in the solution. Table 1.2 shows the effects of the addition of 1 mole of solute to a solution with regard to each of the colligative properties. All of the colligative properties covary in a linear manner with the solvent mole fraction. By measuring any one of these characteristics (e.g., freezing point), one can determine quantitatively any other (e.g., osmotic concentration).

It is very inconvenient to always be calculating the mole fraction of either the solvent or the solute. This is not a value that we tend to think of or measure. For the sake of convenience, we can simplify this unit substantially.

Table 1.2. The effects of the addition of 1 mole of solute to a solution with regard to each of the colligative properties

Concentration (moles/kg H_2O)	*Glucose*		*NaCl*	
	Concentration of effective particles	*Activity as % of perfect solvent*	*Concentration of effective particles*	*Activity as % of perfect solvent*
0.1	0.096	96	0.192	96
0.5	0.542	108	0.937	94
1.0	1.158	116	1.885	94

The expression for the mole fraction of solute is

$$\text{Mole fraction of solute} = \frac{\text{Mole fraction of solute}}{\text{Mole fraction of solute} + \text{More fraction of solvent}} \tag{1.9}$$

As we can see from the calculation shown in Eqn 1.8, in biologically relevant solutions the mole fraction of solutes is very low in proportion, usually <2%. Therefore, the mole fraction of solvent in such solutions is a very large number relative to the mole fraction of solute. We can simplify Eqn 1.9; therefore, by leaving the mole fraction of solute out of the denominator. We are left with the equation

$$\text{Mole fraction of solute} \approx \frac{\text{Mole fraction of solute}}{\text{Mole fraction of solvent}} \tag{1.10}$$

The expression to the right of the equality symbol is equivalent to the number of moles of solute per mole of solvent; that is, the molar concentration. For practical purposes, therefore, we can consider the effects of solutes on the colligative properties of water to be linearly related to the molar concentration of the solutes in the solution.

1.5 The activity coefficient

In the previous section, we explained and pointed out that the colligative properties of water (including osmotic concentration) can be quite accurately determined if one knows the molar fraction of solute present in the aqueous solution. The mathematical and physical underpinnings of our calculations were based on the concept of a perfect solute. By this we mean that the solute must be perfect in the same way that we refer to perfect gases, that is, the particles must not interact with each other in any way. I am sorry to inform you that perfection is hard to come by. In the real world, such assumptions about perfect solutes may not be entirely met.

The most obvious examples of this are found among the electrolytes. For example, if we add one-tenth of a mole of NaCl to a liter of water, the first thing that happens is that the NaCl dissociates into its constituent ions, namely Na^+ and Cl^-. As a result, one-tenth of a mole of NaCl yields two-tenths moles of particles. Each of these ions acts as a separate and effective solute. By comparison, if we dissolve one-tenth mole of sucrose in a kilogram of water, the sugar dissolves but the molecules remain intact.

One-tenth mole of sucrose yields one-tenth mole of solute particles when dissolved in water.

Electrolytes yield additional particles depending on chemical structure. NaCl and KCl yield two particles in solution. $CaCl_2$ yields three, one calcium ion and two chloride ions. The important point is not the number of moles of substance added to the solution but rather the number of discrete particles produced upon interaction with the solvent.

Most electrolytes act as perfect solutes at low concentration (e.g., in a 0.1-M solution). Each ion moves about in its own sphere unaffected by other solute molecules. After all, the solutes constitute only about 0.1% of the molecules present while the water molecules constitute 99.9% of the molecules present. At higher concentrations, however, solutes can begin to interact, decreasing their capacity to act as perfect solutes. When NaCl dissolves in water, the following reaction occurs:

$$NaCl \rightleftharpoons Na^+ + Cl^-$$

That is, the salt dissociates into its ionic forms in a reversible manner. At low salt concentration the dissociation is nearly complete. At higher concentrations, the Na^+ and Cl^- have a greater likelihood of interacting with each other and they will sometimes come together again through electrostatic interactions. I am not referring to precipitation of salt which would occur when the salt concentration exceeds the solubility level. Instead, I am referring to interactions of the two ions while both remain in solution. Under these conditions, because of their close association, the two ions act as a single solute particle. Table 1.3 shows the effects of increasing solute concentration on the effective solute activity of the dissolved solutes for glucose and NaCl. It can be seen that when solutes are present in low concentration they act similar to perfect solutes. As the concentration of the solutes increases, their deviation from "perfect" behavior increases. Large organic

Table 1.3. The osmotic effects of increasing solute concentration

Moles added per liter of water	*Osmotic concentration if the solute is glucose*	*Osmotic concentration if the solute is NaCl*
0.028	0.025	0.03
0.2	0.13	0.38
0.4	0.42	0.73
0.6	0.66	1.10
0.8	0.93	1.49
1.0	1.18	1.92
1.2	1.45	2.27

solutes often have an effective concentration in excess of their true concentration (see Chapter 3 for a discussion of the interaction of water with organic molecules) while electrolytes can have a lower effective concentration due to the effects of ionic interactions. The true concentration of a solute as indicated in Table 1.3 is the actual amount added to the water present. The effective concentration is the concentration of particles contributing to the mole fraction of solute. The ratio of these two values is the activity coefficient.

When speaking of the "activity of a solute," we are referring to the capacity of the solute to act as a perfect solute, that is, its capacity to lower the vapor pressure of the solvent, raise the boiling point, etc. We can similarly use the term the "activity of water" to refer to the degree to which the solvent in a solution differs from pure water.

1.6 The activity of water

In this chapter, we have discussed the fact that the presence of solutes modifies the properties of water. From a chemical viewpoint, the presence of solutes lowers the Gibbs free energy of the water (solvent) molecules. This lowering of free energy affects the colligative properties of the water. For example, solutes lower the vapor pressure above the solution, and raise the boiling point. Similarly, the capacity of water molecules to diffuse through the solution is reduced in proportion to the reduction in the mole fraction of water. A reduction in free energy of the water affects the driving force for diffusion.

It is cumbersome to refer constantly to the Gibbs free energy of the solvent in addressing issues of diffusion, solubility, and osmosis. Therefore, physiologists employ the term "the activity of water" to refer to the capacity of water to diffuse, create vapor pressure, and dissolve solutes. The activity of water in a solution is compared to pure water. Addition of solutes lowers the activity of water. This lowering is precisely and quantitatively related to the reduction in the mole fraction of water in the solution. Although, as we have seen, solutes in the real world may not always act as perfect solutes, for dilute solutions this reduction parallels the effective molar concentration of solutes. The concept of the activity of water as the driving force for osmosis will be the topic of our next chapter.

Suggested additional readings

Blandamer, M.J., J.B.F.N. Engberts, P.T. Gleeson & J.C.R. Reis (2005) Activity of water in aqueous systems: a frequently neglected property. *Chem. Soc. Rev.* 34:440–458.

Franks, F. (1983) *Water.* Royal Society of Chemistry, London.
Kirschner, L.B. (1991) Water and ions. In: *Environmental and Metabolic Animal Physiology,* C.L Prosser, ed. Wiley Liss, New York.
Price, N.C. & R.A. Dwek (1974) *Principles and Problems in Physical Chemistry for Biochemists.* Clarendon Press, Oxford.

2 Osmosis

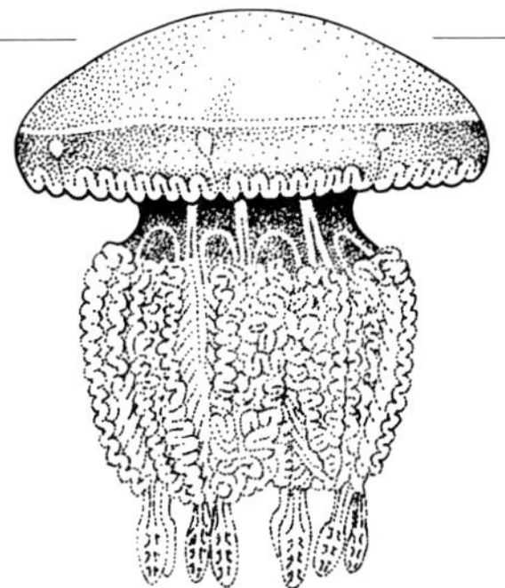

2.1 The phenomenon of osmosis

Consider the following situation. We place a shallow dish on the table in front of us and we pour a 1 molar (1.0 M) solution of sucrose in water into that dish. Next, a more dilute aqueous solution of sucrose (0.1 M) is poured carefully into the dish such that the two solutions are layered one on top of the other. The more dilute solution will float on top of the more concentrated one, because it is less dense. You know this to be the case because you have seen that when you add sugar (sucrose) to your glass of ice tea, the concentrated sucrose solution accumulates at the bottom of the glass until you stir it. In our shallow dish, both water and sucrose molecules are moving about because of kinetic energy, that is, they are diffusing randomly.

After a very long period of time, the gradients in our dish will disappear. The two solutions will come into equilibrium and will have identical sucrose concentrations. This will occur because the effects of diffusion will destroy the gradients, causing the fluids to become uniform throughout the dish. Let us take a closer look to see why this is the case.

First, let us examine what is happening with the sucrose molecules. The concentrated solution has more sucrose molecules in any given volume than does the dilute solution. In fact, as the concentrated solution has a concentration of 1 mol/l and the dilute solution has a concentration of 0.1 mol/l, initially, there are exactly 10 times as many sucrose molecules per milliliter as in the more concentrated solution. At the interface between the two solutions, therefore, there are 10 times more sucrose molecules on one side than on the other side (Fig. 2.1). All the sucrose molecules are diffusing randomly, driven by the kinetic energy they possess. Figure 2.2 shows that for purely

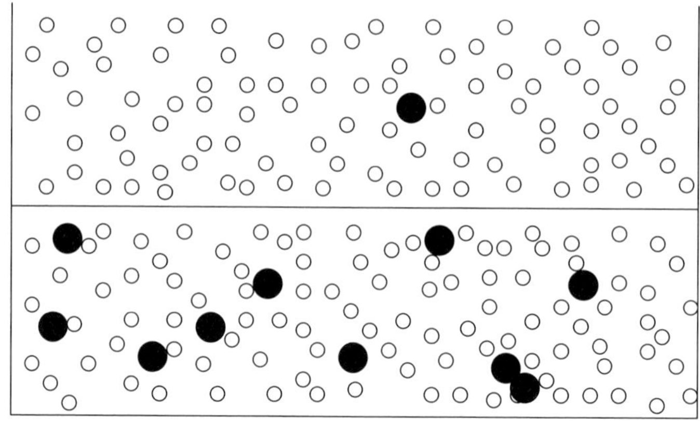

Fig. 2.1. An illustration of initial conditions when a dilute sucrose solution is layered on top of a more concentrated one. The small open spheres are water molecules, the larger darker spheres are sucrose molecules.

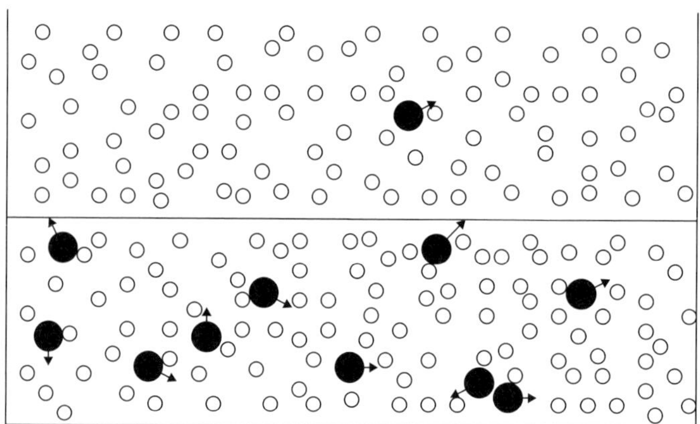

Fig. 2.2. As the sucrose molecules undergo random movements, the solution with the most sucrose molecules (the lower solution) would also have the most molecules randomly crossing the line of demarcation to the more dilute solution.

statistical reasons, more sucrose molecules will diffuse upward than downward under these conditions. Sucrose shows a net diffusion from areas of high concentration to areas of low concentration. This occurs because there are more sucrose molecules near the interface on the side with high concentration. Therefore, for purely statistical reasons, more sucrose molecules will cross the interface going from high to low concentration than in the opposite direction. This occurs solely because of the cumulative effects of random motions and not because of the fact that sucrose "loves" water, or is repelled by other sucrose molecules, or "knows where it should be." The movements are strictly random.

When we consider the water molecules, the same principles apply. In the case of water, the concentration gradient is reversed. The concentration of water is higher in the more dilute fluid. As discussed in Chapter 1, we can refer to the concentration of water as the activity of water, meaning the capacity of water to engage in chemical reactions, to dissolve solutes, and to diffuse freely. In our dish containing two fluids, water molecules show a net diffusion from the more dilute sucrose solution to the more concentrated one as a result of water moving down its concentration (activity) gradient. For purely statistical and physical reasons, therefore, sucrose displays net upward diffusion, and water displays net downward diffusion in our dish.

Now consider a situation in which the two solutions are configured as described above but they are separated by a semipermeable membrane. A semipermeable membrane is one that is permeable to some substances but not to others. This is a common characteristic of biologic membranes. In this case, let the membrane be permeable to water but not to sucrose.

Under these circumstances, the sucrose cannot diffuse down its activity gradient because it cannot pass through the membrane. Water molecules can, however, diffuse down their activity gradient through the membrane. This would result in water moving from the dilute sucrose solution to the concentrated one. This is called *osmosis*. Osmosis is a special case of diffusion, used to describe the diffusion of water down its activity gradient through a semipermeable membrane. It is important to recognize that the water is not "attracted by the more concentrated solution." Instead, water is simply diffusing down its activity gradient.

Unlike the example above of water diffusing vertically in a shallow dish, we are usually more interested in the movement of water into and out of cells, tissues, or even entire animals. This is because osmosis is the process that drives water through cell membranes and across the epithelial surface of animals. Biologists normally illustrate fluid flow and solute transport in diagrams which show horizontal fluid flow, as illustrated in Figure 2.3. In this figure, we can see that a differential activity of water across a semipermeable membrane is causing a net movement of water from left to right. The rate of water movement in this system is determined by three fundamental characteristics of the system: (1) the permeability of the membrane to water, (2) the size of the membrane (specifically its surface area), and (3) the difference in the activity of water across the membrane. Biologists describe this relationship with a formula

$$J = D(A)\,(C_1 - C_2) \tag{2.1}$$

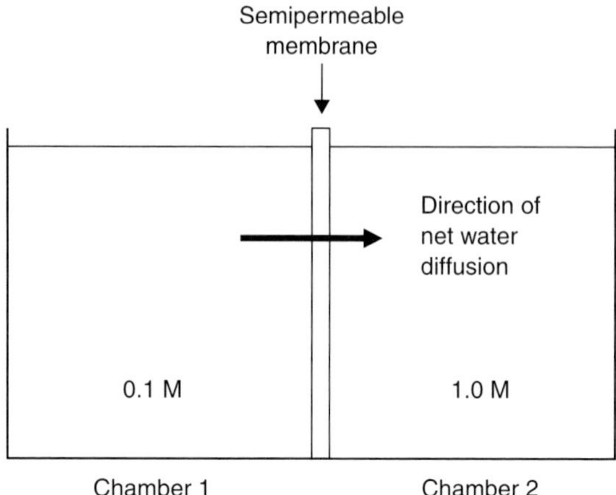

Fig. 2.3. Water moves down its activity gradient, leading to a net diffusion across the semipermeable membrane from left to right.

In this equation, J is the rate of water movement, often referred to as flux. Let us describe the flux in units we can all understand: ml/min. J is, therefore, the rate of water movement across the barrier in ml/min. "A" in Eqn 2.1 is the surface area of the membrane (e.g., cm^2). Clearly, if you have a driving force for moving water across the membrane, the larger the surface area the more water gets moved. Equation 2.1 tells you that the rate of flux is directly proportional to area. C_1 in Eqn 2.1 is the concentration of solutes in compartment 1 and C_2 is the concentration of solutes in compartment 2. For the reasons described in Chapter 1, we express these concentrations in Osmoles (Osm) as these units express the activity of water. This leaves D, the diffusion coefficient. D is a constant for each membrane which describes the inherent permeability of the membrane to water flux. Some barriers are very impermeable to water (e.g., the skin of a snake), while others are highly permeable (e.g., the diaphanous membranes covering a jellyfish). Let us see what units we need to use for D that would allow us to have diffusion coefficients that would be useful for the wide variety of biological barriers we are likely to encounter.

J is expressed in ml/min. As a milliliter is also a cubic centimeter, we can express J as cm^3/min. A is expressed in cm^2. C_1 and C_2 are expressed as Osm, and as we are subtracting Osm from Osm, the units remain Osm. If we adjust Eqn 2.1 to solve for D (do not you love this fancy algebra jargon?) we get

$$D = J/A\ (\text{Osm}) \tag{2.2}$$

Table 2.1. Approximate values for the osmotic permeability of various epithelia

Epithelium	*Osmotic permeability (cm/min Osm)*
Mammalian lung	1.0
Mammalian amnion	0.9
Mammalian colon	0.1
Toad bladder (no ADH)	0.02
Toad bladder (+ADH)	0.4
Mammalian sperm membrane	0.001
Fish gill	0.0003
Fish skin	0.00008
Crab cuticle	0.0006
Lizard skin	0.00001
Snake skin	0.000001
Cuticle (aquatic insect)	0.0004

Expressed in words, the diffusion coefficient describes the rate of water movement across the membrane per Osm of driving force, for a membrane of a given surface area.

If we insert the units in place of symbols we get

$$\mathrm{D} = \frac{(\mathrm{cm}^3)}{\mathrm{cm}^2\,(\mathrm{Osm})} \tag{2.3}$$

At this point the units still make sense. They reflect the cubic centimeters of water that can cross a membrane of a given surface area for each Osm of force.

By canceling units we can simplify this equation to read

$$D = \mathrm{cm/min\ (Osm)} \tag{2.4}$$

Therefore, when you see values of D for various membranes (such as the ones listed in Table 2.1) the units will be cm/min (Osm). Obviously this now makes no sense at all. How can permeability be expressed in cm/min? Just remember when you see this that the real units before canceling are as shown in Eqn 2.3 and the units do make sense if you consider that D is an expression of the inherent permeability of the material in the membrane.

Therefore, let us review what Eqn 2.1 has told us. Each membrane has an intrinsic permeability to water. That permeability is expressed as the diffusion coefficient. The rate of water movement increases in proportion to (i.e., linearly and directly with) the driving force for water movement expressed as the difference in osmotic concentration. The larger the surface area of the membrane the faster the water moves.

Let us use this equation to solve a practical problem in animal osmoregulation. Goldfish have a blood concentration of about 0.3 Osm, yet they live in fresh water, the osmotic concentration of which can be very low, let us say 10 mOsm. As a result, the activity of water is much higher in the external medium than it is in the body fluids of the fish. The skin and, particularly, the gills of goldfish are very permeable to water, leading to the osmotic uptake of water. We can use Eqn 2.1 to determine the rate at which water enters a goldfish by osmosis.

Let us use a goldfish with a body weight of 20 g. The surface area of a roughly cylindrical animal can be described by the following equation:

$$\begin{aligned}\text{Surface area in cm}^2 &= 10\ (\text{weight in grams})^{2/3}\\ &= 10\ (20)^{2/3}\\ &= 10\ (7.4)\\ &= 74\ \text{cm}^2\end{aligned}$$

Therefore, we can estimate the surface area of the fish to be about 74 cm^2. Let us use the permeability coefficient for the skin (8×10^{-5} cm/min Osm), as shown in Table 2.1.

Using Eqn 2.1

$$\begin{aligned}J &= A(D)\,(C_1 - C_2)\\ &= 74\ \text{cm}^2(8 \times 10^{-5}\ \text{cm/min Osm})\ (0.3\ \text{Osm} - 0.01\ \text{Osm})\\ &= 178 \times 10^{-5}\ \text{ml/min}\end{aligned}$$

A microliter is a thousandth of a milliliter, so the water uptake of the goldfish can be expressed as 1.78 µl/min. This means that the rate of uptake over 1 hr would be 60 times that or about 0.1 ml/hr or 2.4 ml/day. Most animals contain about 75% water; hence, a 20-g goldfish contains about 15 g (15 ml) of water. As the total water content of the fish is 15 ml, we can see that the fish gains through osmosis a volume of water equal to about 16% of its total body water per day. This is a rapid but manageable rate of water uptake. To put this in perspective, in a 70 kg (154 lbs) human, this would be equivalent to drinking 11 liters of water a day. The United States Department of Agriculture recommends that this 70-kg person ingest 3.7 liters of water a day, so one can see that goldfish obtain a large amount of water through osmosis, but one which can be managed. The fish's kidneys will produce copious, dilute urine to rid the body of this water.

Let us use Eqn 2.1 to solve a different problem, one that biologists often face when investigating a new experimental system or organ. How would you determine the diffusion coefficient of an unknown membrane?

Figure 2.4 shows an experimental preparation for determining the diffusion coefficient of a flat layer of tissue, in this case a toad bladder. The bladder has been removed from the animal and rinsed out with distilled water. The lumen of the bladder is then filled with distilled water and tied to a glass rod. The sites for fluid exit from the bladder (the ureters and urethra) are tied. The bladder and tube are then weighed on a balance. Let us say that the combined weight of the tube and bladder is 20 g.

The bladder is then placed in a solution that is isosmotic to the blood of the animal. This solution contains a nonpermeant solute, for example, a 0.3-M sucrose solution. The bladder is left in this solution for 20 min. After this, the bladder is removed from the sucrose solution and the external surface of the bladder is blotted dry with a paper tissue. The bladder and tube are weighed again. The new weight is found to be 18 g. The bladder lost weight over this 20-min period due to the osmotic movement of water from

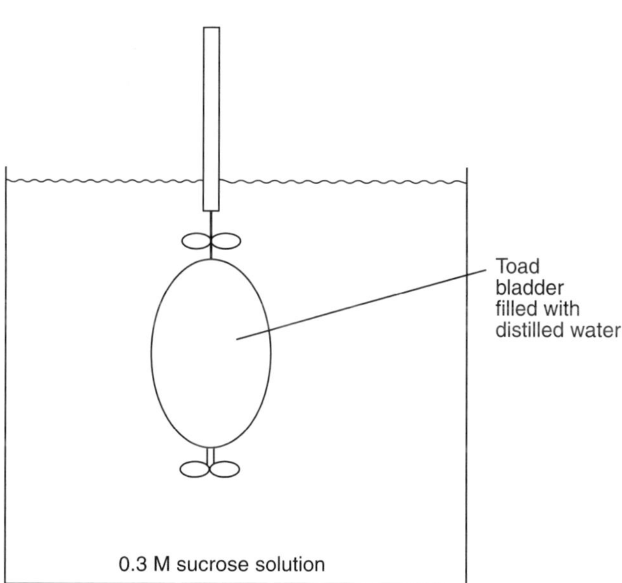

Fig. 2.4. A practical method for estimating the osmotic permeability of an epithelium. Here, a toad bladder is ligated so that any water movement must occur across the epithelium. By establishing a known gradient, the rate of water movement in response to that gradient can be established by weighing the bladder, immersing it in the sucrose solution for a defined period of time, and then reweighing it.

the fluid with a high activity of water (the distilled water) to the solution with a lower activity (the 1-M sucrose solution).

We can now dissect the bladder, cutting it open and measuring the surface area of the bladder. Let us say we find the bladder to have a surface area of 12 cm^2. We can now plug these values into Eqn 2.4

$$J = AD\,dc$$
$$0.1\ \text{ml/min} = 12\ \text{cm}^2\ (D)\ 0.3\ \text{Osm}$$

We rearrange this to solve for D.

$$D = 0.1\ \text{cm}^3/\text{min} \times 12\ \text{cm}^2\ (0.3\ \text{Osm})$$
$$D = 0.36\ \text{cm/min}$$

The diffusion coefficients shown in Table 2.1 were determined in this manner, either by measuring water flow across isolated epithelia or by measuring water uptake into cells and determining volume changes.

It can be seen, therefore, that if you know the diffusion coefficient for a given membrane, you can estimate the rate of water flow across that membrane, assuming that you can estimate surface area and osmotic gradient. Alternatively, using isolated tissues or membranes and precisely controlled osmotic gradients, you can measure water flux and determine the diffusion gradient. Both approaches are used and will be of great use to us when we discuss specific tissues in chapters to come.

2.2 The reflection coefficient

Throughout the above discussions and calculations, we have worked with the assumption that the membrane separating the two solutions is a perfect semipermeable membrane, meaning that it possesses permeability for water but is completely impermeable to the solute. The reader is old enough to know by now that perfection is vanishingly rare. Most membranes have a finite permeability to solutes. What then happens if the membrane is permeable to the solute?

Figure 2.5 illustrates the conditions we have discussed previously, namely, ones in which two solutions with different osmotic concentrations in compartments 1 and 2 are separated by a semipermeable membrane. The difference in this figure is that I have graphed the activity of water in solutions below the chambers. It can be seen that a steep gradient for the activity of water exists right at the membrane. The region in which the greatest net diffusion occurs, due to the gradient in the activity of water, is therefore exactly superimposed on the membrane. As a result, water moves down its activity gradient across the membrane and we have osmosis.

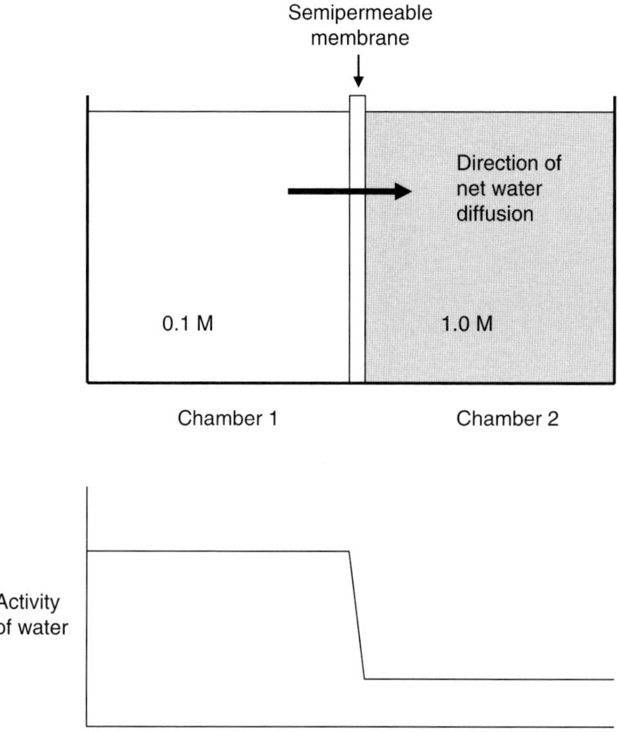

Fig. 2.5. Under initial conditions, a steep gradient for the activity of water lies right at the membrane.

The reader can appreciate that as water moves from left to right in Figure 2.5, the solute becomes more dilute in Chamber 2, resulting in a gradual slowing down of the rate of osmosis. This occurs because the movement of water from left to right across the membrane dilutes the differences in concentration. It is important to note, however, that the location of the steepest part of the gradient does not move, it is always located right at the membrane.

Let us now consider a situation where the membrane is permeable not only to water, but also to the solute contained in Chamber 2. This situation is illustrated in Figures 2.5 and 2.6. The initial conditions are identical to those depicted in Figure 2.5, namely, the steep area of the gradient of water activity is located at the membrane. Shortly thereafter, however, the movement of solute across the membrane has caused the steep part of the gradient to move to the left, and it is now located inside Chamber 1. As a result, there is still a zone of nonequilibrium where water shows net diffusion,

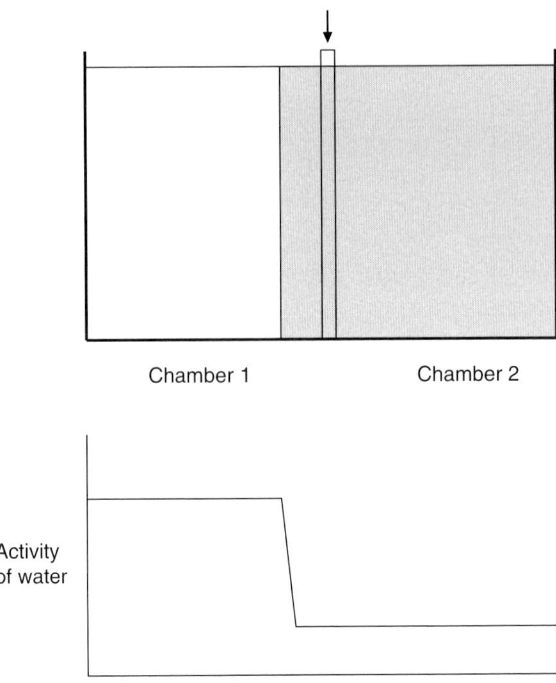

Fig. 2.6. If the solute can penetrate the membrane, then the steep gradient for the activity of water moves away from the membrane and osmotic flow across the membrane ceases.

but this zone is no longer located at the membrane. The membrane, in fact, no longer has an activity gradient across it. For this reason, net water movement across the membrane stops. As we can see, if the semipermeable membrane is permeable to a solute, then that solute is incapable of generating osmotic water movement across the membrane.

How can we deal with this situation quantitatively? Biologists have chosen to address this problem by introducing a new factor in Eqn 2.1, namely, the reflection coefficient. The reflection coefficient is inserted in Eqn 2.1 as follows:

$$J = A(D)\, rr(C_2 - C_1) \tag{2.5}$$

The term, *rr*, is called reflection coefficient because you can think of it as a number indicating the degree to which a solute is reflected back by the membrane. In other words, a solute, which is fully reflected and cannot pass through the membrane at all, has a reflection coefficient of 1. A fully permeable solute has a reflection coefficient of 0.

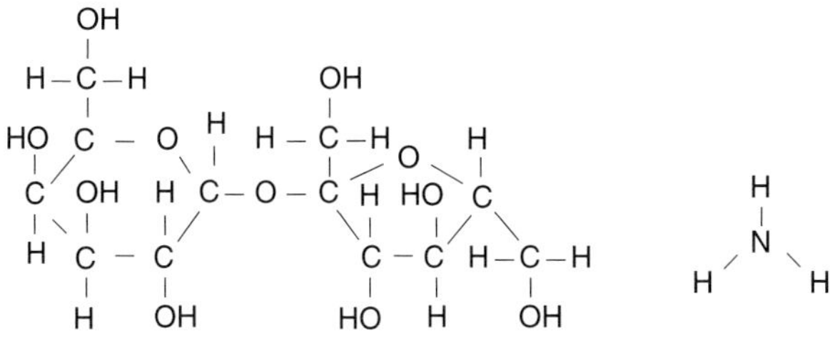

Fig. 2.7. Note the difference in size between an ammonium ion (molecular weight 18) and a molecule of sucrose (molecular weight 342).

If a solute has a reflection coefficient of 1, we can ignore it in Eqn 2.5 because multiplying by 1.0 has no effect on the flux rate (*J*). This is what we did in Eqn 2.1 when we first introduced it, we ignored *rr* because we had set it to 1.0 by declaring that the semipermeable membrane was impermeable to the solute. Conversely, if the value of *rr* is 0, it has a profound effect on Eqn 2.5, causing the flux (*J*) to always be 0. This occurs because a permeable solute can generate no osmotic force due to its own diffusion across the membrane.

You can, therefore, see that the reflection coefficient describes an important osmotic property of a membrane. The situation can even get more complicated. As the reflection coefficient represents the properties of a specific membrane with regard to a specific solute, that number can vary with the solute being considered. Let me give an example. One can imagine a membrane that is relatively permeable to ammonia. Ammonia is a relatively small molecule and most biologic membranes have some finite permeability to ammonia. In contrast, sucrose is a relatively large molecule. It is a disaccharide and unless it is broken down into its constituent sugars, most membranes will be impermeable to sucrose. Figure 2.7 shows the relative sizes of molecules of ammonia and sucrose.

Consider, therefore, a membrane that is separating two solutions as shown in Figure 2.5. Chamber 1 contains pure distilled water while in Chamber 2 there is a solution containing ammonia and sucrose dissolved in water. How fast would water move from Chamber 1 to Chamber 2 under these circumstances? We know that the membrane has a finite permeability to ammonia so let us say that the reflection coefficient for that solute is 0.4. We also know that the membrane is quite impermeable to sucrose so let us say that the reflection coefficient for that solute is 1.0. We then need to modify Eqn 2.5 to take into account the rate of water movement driven by each solute individually.

$$J = A(D)\, rr\left(C_{\text{ammonia}_2} - C_{\text{ammonia}_1}\right) + A(D)\, rr\left(C_{\text{sucrose}_2} - C_{\text{sucrose}_1}\right) \quad (2.6)$$

where J, A, and D are the flux of water, area of the membrane, and diffusion coefficient of the membrane, respectively. C_{ammonia_2} is the concentration in Osm of ammonia in Chamber 2 and C_{ammonia_1} is the concentration in Osm of ammonia in Chamber 1. C_{sucrose_2} and C_{sucrose_1} represent the concentrations of sucrose in these two chambers.

In the example we are using we can substitute the reflection coefficients for each solute into Eqn 2.6, yielding

$$J = A(D)\, 0.4\left(C_{\text{ammonia}_2} - C_{\text{ammonia}_1}\right) + A(D)\, 1.0\left(C_{\text{sucrose}_2} - C_{\text{sucrose}_1}\right) \quad (2.7)$$

This equation states that, in the initial condition, each solute generates a gradient of water activity across the membrane, and is capable of producing osmotic water flow across a semipermeable membrane. The reflection coefficient differs for each solute, however, due to its particular permeability through this membrane. Therefore, the osmotic effect of an ammonium gradient is only 0.4 times as great as a similar sucrose gradient in moving water across this membrane.

There is an additional point I should make, namely, that both ammonia and sucrose are driving osmotic water movement across the same membrane with the same surface area. Therefore, Eqn 2.7 could be simplified to yield

$$J = A(D) \left\lfloor 0.4\left(C_{\text{ammonia}_2} - C_{\text{ammonia}_1}\right) + 1.0\left(C_{\text{sucrose}_2} - C_{\text{sucrose}_1}\right)\right\rfloor \quad (2.8)$$

One could imagine a slightly more complicated situation in which the gradient for ammonia was oriented in one direction (e.g., with Chamber 1 being more concentrated than Chamber 2) and the concentration of sucrose being exactly opposite (Fig. 2.8). In this situation, if the two solutes were present in the two solutions in equal osmotic concentrations, they would affect the activity of water identically. However, because of the differential permeabilities of the solutes, we would see an osmotic flow of water across the membrane as described by the following equation:

$$J = A(D) \left\lfloor 0.4\left(C_{\text{ammonia}_2} - C_{\text{ammonia}_1}\right) - 1.0\left(C_{\text{sucrose}_2} - C_{\text{sucrose}_1}\right)\right\rfloor \quad (2.9)$$

The change in sign between Eqns 2.8 and 2.9 reflects the differences in the directions in which the osmotic forces move water. In the situation described in Figure 2.9 the forces oppose each other and, therefore, the two values must be subtracted from each other. We see a flow because although the activity of water is identical on both sides, the reflection coefficients of the two solutes are not identical.

At this point you may be starting to feel that this is getting hopelessly complicated. You may well be right! Think about the fact that most real-world osmotic problems in biology involve calculations of water movements across membranes that are in contact with blood or seawater. Both are very complex fluids. Seawater has osmotically active concentrations of sodium, potassium, chloride, sulfate, carbonate, and hydrogen ions. And those are only the inorganic compounds! Biologic membranes would have a distinct reflection coefficient for each of these solutes. How can we deal with this complexity and still produce meaningful results of use to us in biologic situations?

We can obtain biologically useful information by measuring osmosis under conditions that are as natural as possible. Let us return to the situation we first looked at in Figure 2.4, namely, the rate of water movement across the bladder epithelium of a toad. In our first approach to that example, we measured the osmotic permeability of the bladder by placing a sucrose solution on one side and distilled water on the other. What if we use a more biologically accurate condition, namely, distilled water inside the bladder and toad's blood (or a saline solution similar to that of blood) as the outer solution? If we had to calculate the rate of water movement across the bladder using the reflection coefficients for all of the solutes in the blood (sodium, chloride, bicarbonate, glucose, etc.), we would face a horrendous mathematical problem. Instead, we can simply do the experiment directly. We place water inside the bladder and toad's blood (or saline) outside. We measure the rate of water movement and calculate the diffusion coefficient for this membrane under this specific set of conditions. This is an excellent experiment and it provides us with a very useful value. After all, these are the osmotic conditions that apply in the animal in its normal physiological state. That is generally what we want to know: how does this animal or tissue function in its normal environment? By determining these values empirically, we can avoid worrying about the effects of each solute individually.

As a savvy consumer of tables of diffusion coefficients, therefore, you need to be aware that the reported values of D can vary with the types of fluid on either side. Generally, biologists will have determined the value of D using solutions that approach as nearly as possible in the natural state, and in normal physiological environment, of the membrane.

2.3 Osmotic pressure

Often, when reading the literature associated with osmotic regulation, one comes upon the term osmotic pressure. This is usually encountered in older literature and in the field of animal osmoregulation the term has largely gone out of use. Let us examine how the term came into being and explore what pressure has to do with osmotic concentration.

Imagine a situation as depicted in Figure 2.8a in which a solution with 1 Osm concentration of an impermeant solute is separated by a semipermeable membrane from a solution with a concentration of 0.1 M. You are now fully familiar with what would happen. Water would move down its activity gradient from left to right. Now let us assume that the walls of the chamber on the right cannot expand outwardly, but the level of fluid in Chamber 2 can rise. The movement of water down its activity gradient would indeed provide sufficient potential energy to lift water. At some point, the system would come into equilibrium, as illustrated in Figure 2.8b.

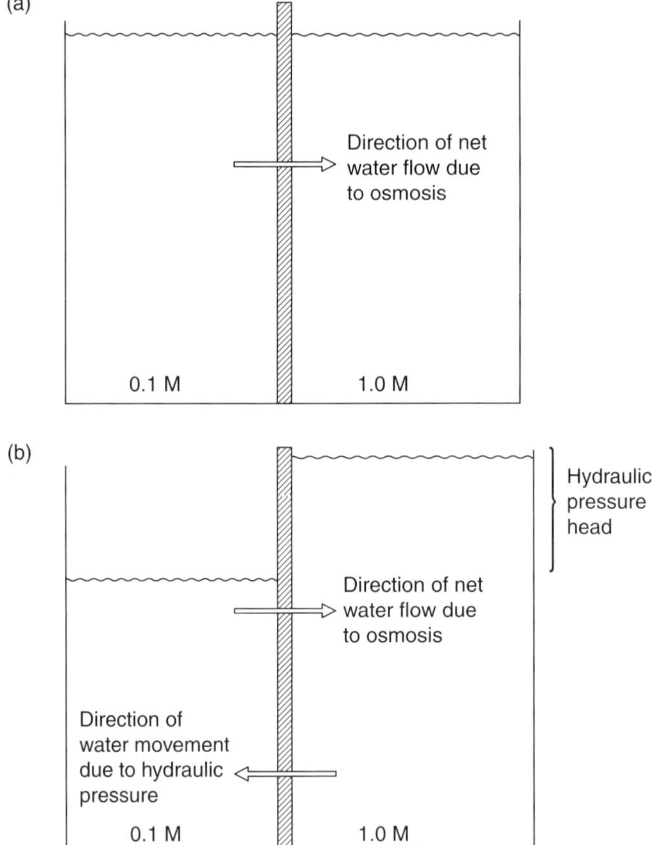

Fig. 2.8. (a and b) The energy stored in the form of a difference in the activity of water can be used to produce a net diffusion of water and a hydrostatic pressure head.

It would come to this equilibrium because the accumulation and, thus, lifting of water in Chamber 2 would create a hydraulic pressure that would drive water back through the membrane from Chamber 2 to Chamber 1. In other words, the hydraulic pressure generated by the difference in the height of the two solutions on either side of the membrane (the hydraulic pressure head) would be equal and opposite in effect to the net diffusion of water from Chamber 1 to Chamber 2.

It is important to recognize that we are dealing here with two distinct phenomena. One is osmosis, which is a process that requires a semipermeable membrane. The second is fluid flow driven by hydraulic pressure. This does not require a membrane (note how water flows through the tap on your sink, or down a river). Hydraulic pressure can move water through a membrane, presumably through various aqueous pores. We can express the effects of the hydraulic pressure across a membrane relatively simply by an equation:

$$J = A\,Lp\,(dP)$$

where J is the flux of water driven by hydraulic pressure, A is the surface area of the membrane, Lp is the hydraulic permeability of the membrane, and dP is the hydraulic pressure gradient across the membrane.

At the equilibrium depicted in Fig. 2.8b, no net water movement occurs because the net osmotic flow equals the net hydraulic flow. That is,

$$J = 0$$

when

$$J = A(D)\,(C_2 - C_1) - A\,Lp\,(dP) = 0$$

It can be seen that the movement of water from left to right in Fig. 2.8b is due to the osmotic gradient, and the movement of water from right to left is due to hydraulic flow. The hydraulic flow is directly proportional to the difference in height of the water level in the two solutions. Therefore, the difference in the osmotic concentration of the two solutions is proportional to the difference in height of the two solutions at equilibrium. This system can therefore be used to measure the osmotic concentration of solutions. If you can measure the difference in height of the fluids in the two chambers, you can get an estimate of the osmotic concentrations of the two fluids. If one solution has known osmotic concentration (e.g., a standard solution that you prepared), this method can tell you the concentration of the other, unknown solution. This method was used in older experiments to measure osmotic concentration (e.g., by van't Hoff in 1887). As described in Chapter 1, there are now much easier and faster ways of measuring the osmotic concentration of solutions. Nonetheless, it is important to be aware that hydraulic pressure can also drive water through a membrane.

The important points at this time are (1) that water can be moved through a membrane either by an osmotic gradient or by a hydraulic pressure gradient, and (2) that these forces can be additive (or subtractive) depending on the circumstances.

In animal systems, we usually do not worry about hydraulic pressures. In plants, by contrast, hydraulic forces generated by osmosis are of critical importance. Even though this book is about animal osmoregulation, let us take a quick look at how plants generate hydraulic forces from osmosis.

Imagine the situation I described in Figure 2.9a, but now with the circumstance that there is a piston on top of Chamber 2. As fluid flows from Chamber 1 to Chamber 2, the piston would be lifted (Fig. 2.9b). Alternatively, downward pressure can be applied to this piston to resist fluid flow through the membrane from Chamber 1 to Chamber 2. If we put enough pressure on this piston, we can resist the expansion of Chamber 2. Under these conditions, the system is again in equilibrium (Fig. 2.9c). There is no net fluid flow when the osmotic force driving water from Chamber 1 to Chamber 2 is opposed by the hydraulic force applied by the piston. This is precisely the situation in plant cells. Plant cells are surrounded by cell walls made up of cellulose. The plant cells are filled with cytoplasm (a solution containing many osmotically active solutes) while the extracellular fluids are much more dilute (being composed mostly of water pulled up from the roots). Water flows by osmosis into the plant cells which then swell. Eventually their cell membranes bump up against the cell walls and can no longer expand. This situation generates hydraulic pressure inside the cell until that pressure matches the osmotic force driving water in. The result in plants is called turgor pressure, the positive hydraulic pressure inside plant cells that gives the plant stiffness and volume.

There are few situations in animals in which hydraulic forces are important in driving fluid movement. As we will discuss in Chapter 10, the kidney of higher vertebrates uses blood pressure to drive fluid through the walls of the glomerulus, thereby producing the primary urine. We will examine this process in more detail when we discuss kidney function.

There is a final point that needs to be made about the capacity of osmosis to generate pressure. When an animal cell is placed in a solution more dilute than the fluids in the cytoplasm, water will move by osmosis across the plasma membrane and into the cell. This causes the cell to expand and swell, that is, to expand in volume. The cell cannot swell indefinitely, because the surrounding cell membrane can only expand a small amount before it ruptures. Is the force generated by osmotic water flow really sufficient to blow a cell apart? Let us take a look.

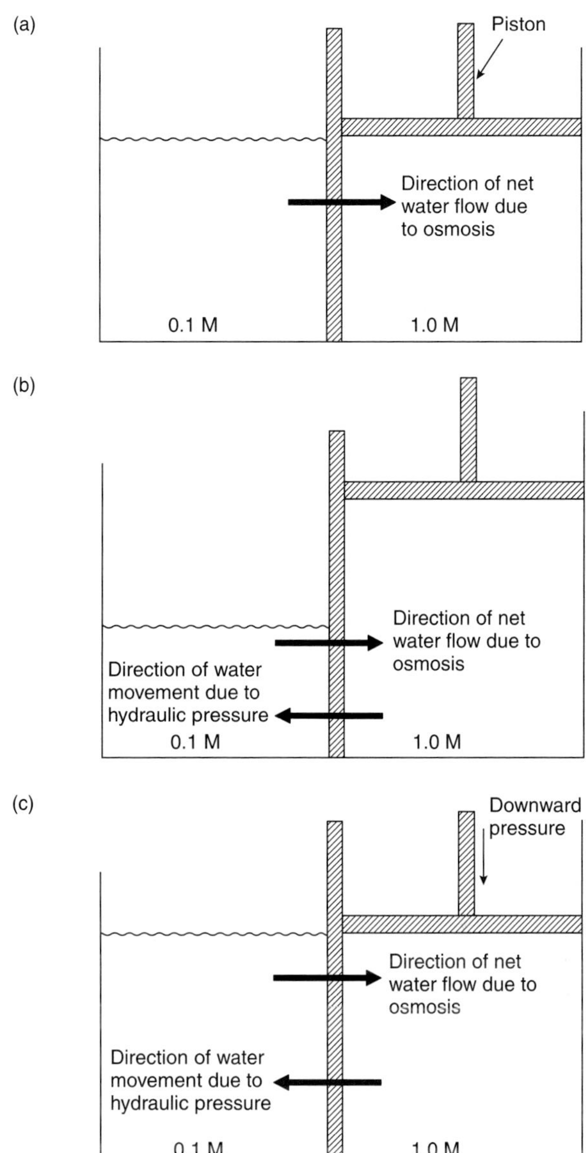

Fig. 2.9. (a–c) Application of pressure can be used to produce a gradient for the activity of water across a semipermeable membrane.

The pressure that can be generated by osmosis is proportional to the difference in the osmotic concentration of solutes on opposite sides of the membrane. As we discussed above, one can measure directly the pressure produced under different osmotic conditions. It turns out that a gradient of

1 Osm of solute across a membrane impermeant to that solute will generate a pressure of 22.4 atm. That sounds like a lot of pressure, but frankly most of us are not used to dealing with pressure in atmospheres. Let us pick some units with which we are more familiar.

In the United States and Great Britain, tire (or tyre) pressures are measured in pounds per square inch (lbs/sq. in.). In the rest of world, they generally use the pressure unit of bars (equal to the pressure generated by a column of mercury 1-m tall). Your average automobile tire operates with a pressure of about 35 lbs/sq. in.

$$35 \text{ lbs/sq. in.} = 2.41 \text{ bar}$$

I think we can all agree that the pressure in an automobile tire is very high. It is sufficient to keep a tire inflated even with a big automobile pressing down on it. In addition, the walls of the tire are thick rubber, designed to keep that pressure in. If we express that tire pressure in atmospheres it turns out to be 2.4 atm.

Therefore, let us go back to the pressure that a 1 Osm osmotic gradient can generate, namely, 22.4 atm. The pressure that 1 Osm gradient can generate is more than nine times greater than the pressure in your automobile tires! Small wonder then, that osmotic forces can rupture cell membranes that are 100,000 times thinner than a millimeter!

Fortunately animals, whether they be insects, fish, worms, or humans, have many defenses that prevent them from developing these enormous pressure forces. The following chapters in this book will provide insights into these processes and their diversity among animals. A full appreciation of the importance and efficacy of those processes requires an understanding of the deleterious effects of osmotic gradients. These deleterious effects include potentially massive pressure forces, osmotic, and ionic perturbations in cellular compartments, and the large fluxes of water that are an integral part of life for organisms inhabiting many of this planet's diverse osmotic habitats.

Suggested additional readings

Hochachka, P.W. & G.N. Somero (2002) *Biochemical Adaptation*. Oxford University Press, Oxford.

Verkman, A.S. (2000) Water permeability in living cells and complex tissues. *J. Membr. Biol.* 173:73–87.

Withers, P.C. (1992) *Comparative Animal Physiology*. Saunders College Publishing, Forth Worth, TX.

3 The interaction of water with organic solutes

3.1 The effects of polar solutes

Polar solutes such as salts can dissolve in water. Let us take table salt (NaCl) as an example. If present in sufficiently dilute concentration, sodium chloride will dissociate in water into sodium ions (Na^+) and chloride ions (Cl^-). Note that the table salt which was electrically neutral (Do you notice any lightening bolts coming out of your salt shaker?) dissociates into two ions, each of which is charged, but which are neutral overall. The charged nature of the ions has specific effects on the surrounding water molecules. Positive sodium ions tend to influence the alignment of the water molecules around them such that the most statistically likely orientation of the water molecules is with the negative pole of the water molecules oriented toward the sodium ion. Conversely, the most statistically likely orientation of the water molecules around a chloride ion is with the positive pole of the water molecules oriented toward the negatively charged chloride ion (Fig. 3.1). This "shell" of water around charged solutes serves to keep the solute associated with water and dissociated from its polar opposite. For example, NaCl dissolves readily in water into Na^+ and Cl^-, but this salt is essentially insoluble in nonpolar solvents such as benzene and olive oil. The interactions between water and polar solutes, therefore, have important effects both on the structure of water around the solute and on the solubility of the solute in water.

A number of organic solutes also have polar portions that interact with water and/or form hydrogen bonds with surrounding water molecules. Common sugars are a good example. Figure 3.2 shows the molecular structure of glucose. It can be seen that many of the carbons in the ring have a bond between the carbon atom and the hydrogen atom (i.e., a C–H bond) and that same carbon also has a bond to a hydroxyl "–OH" group (i.e., a C–O–H bond). In the C–H bond, the carbon and hydrogen share the electron fairly evenly so this bond is not highly polar. We have already mentioned that the O–H bond, as found in water, is very unbalanced with the oxygen hogging the electron, leaving the hydrogen with a partial positive charge. The C–O–H bonds on glucose, therefore, readily form hydrogen bonds with water, thereby promoting the solubility of glucose and serving to orient water molecules around this polar region. As we will see later in this chapter, molecules with a predominance of C–H bonds are rather

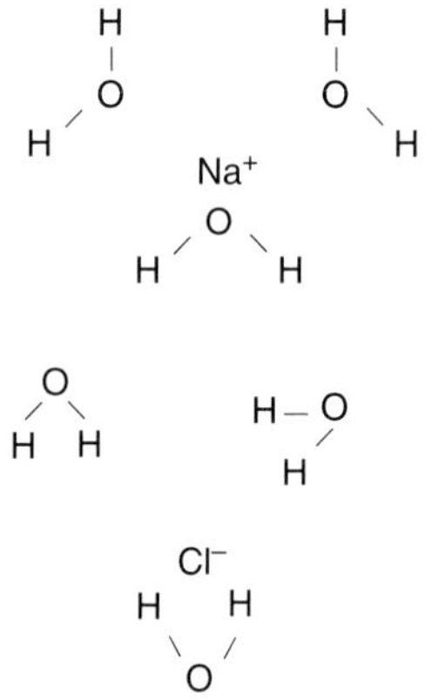

Fig. 3.1. Owing to the charge polarity in the water molecule, the water molecules associated with ionic solutes have a statistically preferred orientation as shown in these diagrams.

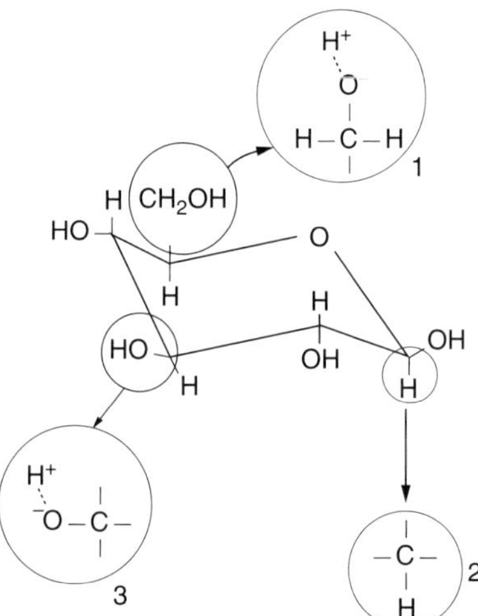

Fig. 3.2. The chair structure of glucose. The hydroxyl groups at each carbon atom can ionize (1, 3), promoting the aqueous solubility of the molecule. The carbon-hydrogen bond (2) does not ionize and therefore does not contribute to aqueous solubility.

insoluble in water while those with lots of C–O–H bonds are more soluble. Sugars, because of these hydroxyl groups, are highly soluble in water and, therefore, can participate actively in the aqueous reactions occurring inside the cell.

Proteins are another class of organic molecules in cells. They are extremely important because proteins are the major structural molecules in the cell. In addition, many proteins act as enzymes, the molecules that facilitate and control almost all the chemical reactions in the cell. All proteins consist of linear chains of amino acids. These amino acids have profound effects on the structure and function of a protein. Let us begin by examining the structure of amino acids and how they might interact with water.

When dissolved in water, amino acids have a number of potential polar regions. Figure 3.3 illustrates the amino acid glycine. You will see that glycine has an amine group (NH_3) and an acid group (COOH), both attached to the same carbon atom. This is a characteristic shared by all amino acids and the ubiquitous presence of the amine and acid groups leads, in fact, to the name amino acid. When dissolved in water, both the amine and the acid groups tend to ionize. The acid group undergoes the reaction (Fig. 3.4).

The amine group undergoes the reaction (Fig. 3.5).

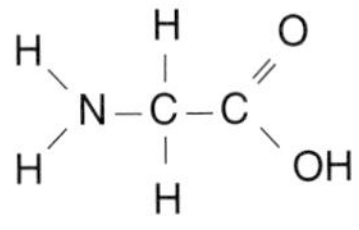

```
H       H   O
 \      |  //
  N  -  C - C
 /      |    \
H       H     OH
```

Fig. 3.3. The structure of the amino acid glycine.

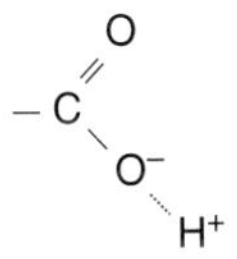

```
       O
      //
  - C
      \
       O⁻
         .
          H⁺
```

Fig. 3.4. The reaction of the carboxylic acid group in aqueous solution.

```
      H
       \    |
    H - N - C -
       /    |
     ⁺H
OH⁻
```

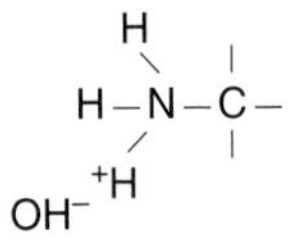

Fig. 3.5. The reaction of the amine group in aqueous solution.

Because one group produces a positive ion and the other a negative ion, the net effect of the two ions is neutral. Although the whole molecule is neutral, glycine is still readily soluble in water because it has locally charged regions that interact with water molecules and promote solubility.

In all amino acids, the amine and acid groups attached to the same carbon atom balance out. The remaining portions of the amino acid (referred to as the side groups) may, however, have portions that can profoundly influence the overall character of the molecule. Take the example of glutamic acid. Glutamic acid has one amine group and two acid groups (Fig. 3.6a). At a pH near neutral, this amino acid will have a net acid character. Glutamic acid, therefore, belongs to the group of amino acids with acidic side groups (Table 3.1). A molecule of glutamate dissolved in water, therefore, interacts with water molecules as expected, orienting the water molecules around it such that the positive poles of the water molecules are more likely to be near the ionized acid group, and with the negative pole of the water molecules oriented near the ionized amine group.

Other amino acids have side groups that contain an additional amine group. At neutral pH, such amino acids would have an overall basic character. An example of such an amino acid is lysine (Table 3.1 and Fig. 3.6b).

Finally, in some amino acids, the types of side groups they possess do not ionize at neutral pH. Alanine (Fig. 3.6c) is an example of such an amino

(a) (b) (c)

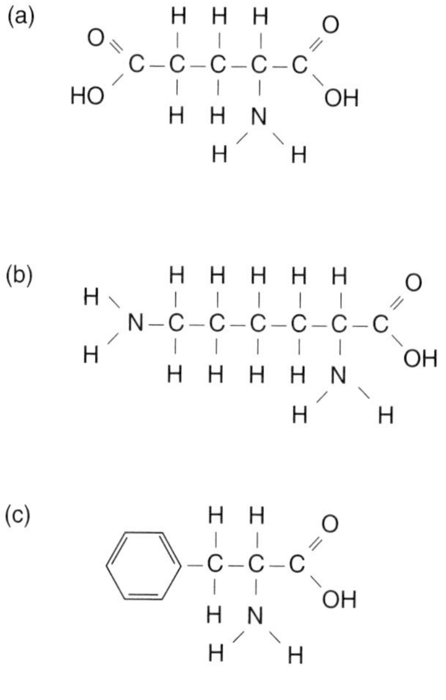

Fig. 3.6. (a) The structure of glutamic acid. (b) The structure of lysine. (c) The structure of phenylalanine.

Table 3.1. List of amino acids on the basis of side chains

Basic side chains	*Acidic side chains*	*Polar side chains*	*Nonpolar side chains*
Lysine	Aspartic acid	Asparagine	Glycine
Arginine	Glutamic acid	Serine	Leucine
Histidine		Threonine	Methionine
		Tyrosine	Tryptophan
		Glutamine	Cysteine
			Phenylalanine
			Isoleucine
			Valine
			Alanine
			Proline

acid. These amino acids are termed neutral amino acids as the amine and acid groups balance out and the side group does not contribute an ion in water. For reasons described below, the types of amino acids in the linear chain of amino acids forming a protein can have a profound impact on that protein's structure and function.

3.2 The primary structure of proteins

Proteins are certainly among the most important molecules in living systems. Proteins form the scaffold of cellular structures. Almost all enzymes, transport moieties, molecular motors, and receptors are proteins. The ability of cells to maintain their form, move, transport, digest, and metabolize depends on the health and well being of the proteins in the cytoplasm.

Proteins consist of linear chains of amino acids. The diverse proteins performing a myriad of functions in cells are differentiated most fundamentally through the types of amino acids of which they are composed, and the order of the amino acids in the linear array.

The linear sequence of a protein is called the protein's "primary structure." It is determined by the gene, which encodes the protein as well as by any editing (snipping and trimming) that may occur after the protein is synthesized. Proteins are synthesized on a cellular organelle termed a ribosome. In the ribosome, the amino acids are attached to each other in a linear manner by means of peptide bonds. The reaction resulting in a peptide bond formation is depicted in Figure 3.7.

You will remember that in water both the amine and acid groups on the amino acid are ionized. When two amino acids are joined through a peptide bond, the amine and acid groups are joined with the result that the two dissimilar charges become electrically neutral. As a result, the amine

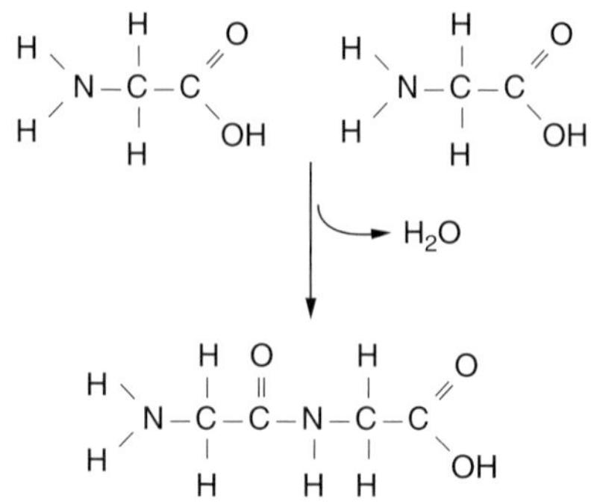

Fig. 3.7. The reaction associated with the formation of the peptide bond.

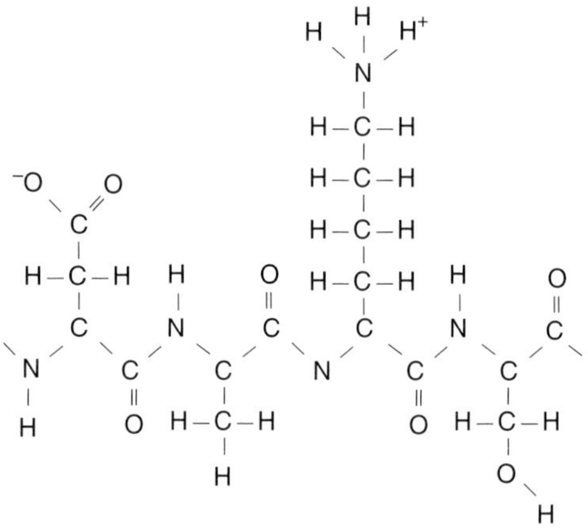

Fig. 3.8. Four amino acids in a peptide chain. Note that the side groups of the amino acid determine the net charge along that portion of the peptide.

and acid groups contributing to the peptide bonds do not contribute to the surface charges on the protein. Instead, the surface charges on the protein is derived exclusively from the side chains. Look back at the illustration of glutamate (Fig. 3.6a). If the amine and acid groups attached to carbon number 1 are tied up in a peptide bond, then the ionic characteristics of the amino acid would be entirely due to the acid group in the side group. Similarly, for lysine (Fig. 3.6b), once the amino acid is in a protein, the side group assures that the net charge on the amino acid is positive. Amino acids with neutral side groups create a region of the protein with no net charge.

The effect of this in a linear array of amino acids (i.e., a protein or peptide) is shown in Figure 3.8. You can see that regions of the protein have various net charges that are determined by the specific amino acids inserted at that location.

3.3 The secondary and tertiary structure of proteins

When a protein is being synthesized as a linear array of amino acids, it comes into contact with the aqueous environment of the cell cytoplasm. At that point, the interactions of the side groups of the amino acids with water molecules take on a huge significance. Acidic amino acids give up their hydrogen ions and tend to build associations with the positive pole of the water molecules. Basic amino acids tend to accept a hydrogen ion and associate with the negative poles of water. As discussed in Chapter 1, however, charged molecules attract not only water but also other ionized molecules with opposite charge. Positively and negatively charged amino acids can therefore develop ionic bonds with each other; bonds that cause the protein chain to bend and adhere to itself. The charges on the amino acids, their effects on water, and their effects on each other serve to produce specific structural associations within the protein. These preferred configurations are referred to as the secondary and tertiary structure of proteins. The term secondary structure refers to regions of the protein in which the associations promote specific protein structures. These include a variety of structural features such as loops, tight hairpin turns, sheet-like structures termed β-pleated sheets, and coiled α-helices. Such structures can be very important in structural proteins, and in the proteins associated with membranes (see below).

The secondary structure of a protein refers to the configuration of a region of the protein. The tertiary structure refers to the overall shape of the entire molecule. As you can imagine, the overall shape of a protein plays a major role in determining its function. A structural protein in the cell is going to be of little use if its interaction with the aqueous solution causes it to fold into a shape not suited for its purpose. Similarly, enzymes must assume a specific tertiary structure in order to produce the active site that interacts with the enzyme's substrate. Inappropriate changes in that tertiary structure would make the enzyme ineffective. The interactions of the side groups on the amino acids with each other, and with their environment determines the shape of the protein (Fig. 3.9).

The secondary and tertiary structures of a protein are, of course, influenced principally by the protein's primary structure. Each protein has a different shape and function because it has a different sequence of amino acids. The tertiary structure of proteins is, however, strongly influenced by the environment in which the protein finds itself. One factor that can affect protein structure is the pH of the surrounding solution. The pH affects the degree of ionization of the side groups in the amino acids. As we have indicated above, the charges on the side chains and their interactions (both

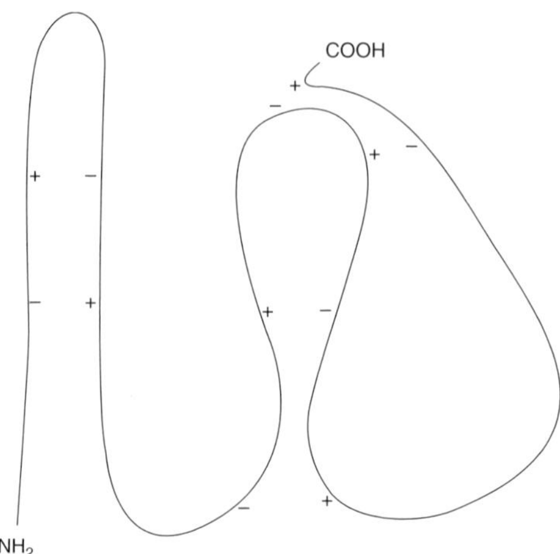

Fig. 3.9. A protein molecule illustrated diagrammatically as a linear molecule. Owing to net charge along its length, portions of the protein form ionic bonds with other oppositely charged regions on the molecule, leading to a preferred configuration of protein folding. The preferred folding pattern can be influenced by the concentration of charged molecules in the fluid surrounding the protein.

attractive in the case of unlike charges and repelling in case of like charges) strongly affect protein shape and stability.

Another important factor is the presence and concentration of strong ions such as salts. Many of the ionized sites on proteins develop strong ionic bonds with free ions in solution such as Na^+, K^+, Ca^{2+}, Cl^-, Mg^{2+}, and SO_4^{2-}. These ions have powerful charges and a small radius, a phenomenon that chemists refer to as high charge density. In an aqueous solution, these ions will have a cluster of water molecules around them. Ionized amino acids on peptides also attract these strong ions, forming ionic bonds. These associations can be transient or longer term, but in either case they can affect the protein structure. The interactions of ions with proteins are inevitable and in many cases beneficial. The role of calcium in regulating cellular function, the transport of sodium and potassium across membranes, and the function of potassium in stabilizing cellular proteins, all require appropriate ionic concentrations in and around the protein. The key term here is "appropriate." Too high concentration of ions can disrupt ionic bonds that keep the protein in its functional configuration. Too low concentration can deprive the protein of ions required for structural or regulatory interactions. Correct cellular function depends on an appropriate ionic milieu around the proteins. As we will see in later chapters, a major

From:
Recycle Bookstore
1066 The Alameda
San Jose , CA 95126
U.S.A.

To:
MYFBAPREP SARASOTA (AVI:
C2906289)
1657 WEST UNIVERSITY
PARKWAY, UNIT A
SARASOTA FL 34243
U.S.A.

FOR CUSTOMS ONLY

Shipping Manifest **Phone:** 9985200899

Sales Order No.: 769869907
AbeBooks Purchase Order No.: 731478909
Processed: July 8, 2026 **Delivery Time:** 5 - 14 business days

aspect of cellular function is the creation and maintenance of an appropriate environment for the cellular machinery through the manipulation of the activity of ions and water.

This brings us to the third parameter that affects protein secondary structure; namely, the activity of water. Many amino acids in proteins have side groups that ionize in water. These ionized sites can either attract oppositely charged amino acids on the same protein, oppositely charged strong ions, or a shell of water molecules. In fact, there is a competition between these various elements, the outcome of which greatly affects protein structure. The activity of water affects the latter two parameters directly. Increasing the activity of water reduces the activity of ions and promotes the likelihood of interactions with water. Decreasing the activity of water has the opposite effect.

It is easy to understand that removing or adding water to the cytoplasm can affect the concentration of an ion (e.g., K^+), and that this would affect proteins the manner similar to adding or removing salt. However, we can show that the activity of water itself affects protein structure. If one adds sucrose to a solution containing an enzyme and substrate, increasing the sucrose concentration to high levels will inhibit enzyme functions. This occurs despite the fact that sucrose does not substantially bind to the protein. Instead, by lowering the activity of water, the increasing sucrose concentration affects the relative concentration of ions present, the interaction of water with charged regions of the protein, and finally even the solubility of the protein itself.

We have ignored for the moment the effects of the neutral amino acids, but let us return to them now. Neutral amino acids are not polar at neutral pH and therefore do not interact very closely with polar water molecules. As a result, the energetically most preferred configuration for the protein is one in which the neutral amino acids are near each other and preferably sequestered away from water. Clusters of neutral amino acids, therefore, often form pockets in the interior of highly folded proteins, placing them close to other neutral amino acids and away from the surrounding water molecules. These associations of neutral amino acids are sometimes referred to as "hydrophobic bonds," meaning that the side chains are held together by their "hydrophobicity," their "fear of water." In truth, these are not actual bonds. Instead, the energy that serves to keep these groups together is provided by the charges on the water, not the lack of charges on the neutral molecules.

We can see that the structure and function of a protein is dependent on the proper control of the activity of water around it. As a result, osmotic regulation is a fundamental and critical function in every living organism.

3.4 The quaternary structure of proteins

Many proteins do not exhibit their full function until they are joined with other proteins into a functional protein complex. Examples of this are the hormone insulin in which the functional unit is two protein chains: hemoglobin, which requires four chains, and glutamate dehydrogenase, which consists of approximately 40 proteins in its functional state. The structure of these protein complexes is termed the quaternary structure of proteins. The ability of these proteins to carry out their functions depends on the precise assembly and appropriate location of each protein in the structure, which in turn depends on the interaction of charged groups on the surface of the proteins. These protein complexes are held together by ionic and hydrophobic interactions, not by covalent chemical bonds. As a result, inappropriate levels of pH, salt concentration, or the activity of water can lead to the disassembly and inactivation of vital protein assemblages throughout the cell.

3.5 The interaction of lipids with water in the cell

The term lipid encompasses molecules with a large range of chemical structures. The common characteristic of the molecules, however, is their low solubility in water. Figure 3.10 shows the structure of oleic acid, a fatty acid found in many organisms. You will note that oleic acid has a long chain of carbon atoms, each of which is attached to one or two hydrogen atoms. These bonds between carbon and hydrogen do not dissociate and are not ionized in water. As a result, water molecules are not attracted to this long, neutral chain of atoms. The first carbon atom of the molecule does possess a carboxylic acid group. The molecule is referred to as a fatty acid because of this acid group. Carboxylic acid does dissociate in water. Nonetheless, the long uncharged chain makes oleic acid very poorly soluble in water.

If lipids are present in water they tend to cluster together and to form accumulations in which water is excluded. You have seen the effects of this in a bottle of salad dressing. The oil and water separate and do not mix. Even if one shakes up the bottle, apparently blending the lipid and aqueous portions, the two phases rapidly separate again. As we will now discuss, this separation of aqueous and hydrophobic phases plays a crucial role in the structure of cellular membranes.

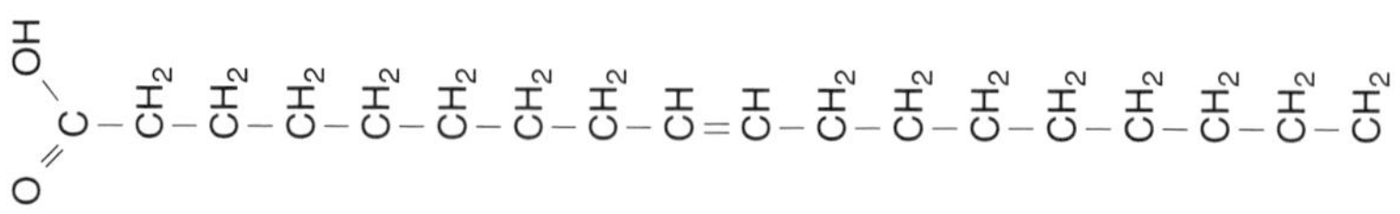

$$\mathrm{HO{-}C({=}O){-}CH_2{-}CH_2{-}CH_2{-}CH_2{-}CH_2{-}CH_2{-}CH_2{-}CH{=}CH{-}CH_2{-}CH_2{-}CH_2{-}CH_2{-}CH_2{-}CH_2{-}CH_2{-}CH_2}$$

Fig. 3.10. The structure of oleic acid.

3.5.1 The structure of membranes

All animals are eukaryotic, meaning that they possess cells with a nucleus. The structures delimiting the cell and separating the nucleus from the remaining cytoplasm are membranes. In addition to defining the external boundaries of the cell and nucleus, cellular membranes are the critical functional units of several cell organelles including mitochondria, the endoplasmic reticulum, the Golgi apparatus, and lysosomes. In considering the structure and function of cells, therefore, it is essential to consider how membranes react with water and how cellular function can be affected by osmotic parameters.

In 1972, Singer and Nicholson described their Fluid Mosaic Model of membrane structure (Fig. 3.11). This model proposes that the membrane consists of a lipid bilayer which serves as a two-dimensional "sea" in which proteins are floating. More recently, this model has been updated a bit, but the major modifications involve a deeper understanding of protein structure within the membrane and a realization that the density of proteins in the membrane is very high (Engelman, 2005) (Fig. 3.12). A better analogy might be, therefore, that proteins are boats afloat in a crowded harbor rather than in a vast "sea" of lipid.

In the previous section, we discussed the fact that lipids are not very soluble in water and tend to partition themselves away from aqueous regions. The lipids in membranes must be rather unusually differentiated, therefore, as membranes surround the cell and have large areas of contact with the aqueous environment. In fact, most membrane lipids do have an unusual makeup in that they are diglycerides. Figure 3.13 shows a single lipid molecule which is typical of the types found in membranes. This lipid is a diglyceride, meaning that it has a glycerol backbone with two fatty

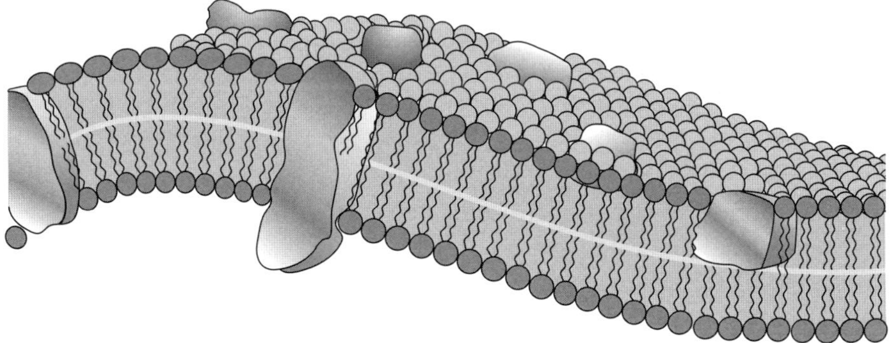

Fig. 3.11. The fluid mosaic model of membranes. (Redrawn from Engelmann, 2005.)

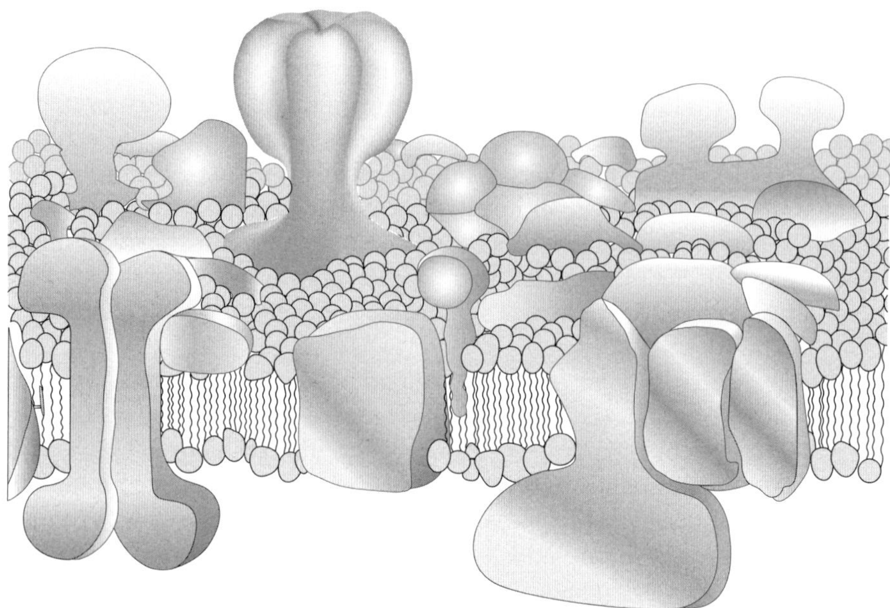

Fig. 3.12. A more modern version of the fluid mosaic model. (Redrawn from Engelmann, 2005.)

acid chains attached to it. The molecule is usually described as having a head region and two tails. The two "tails" are the nonpolar fatty acids that are highly hydrophobic. The "head" region usually consists of a molecular structure capable of being ionized in water, that is, a polar group. In Figure 3.13, the head group is phosphatidylcholine. The precise chemical structure of the head region can vary substantially in various membranes but the head region is always highly polar.

The diglycerides in a membrane are not randomly arranged. In fact they assume a very specific orientation in water. The lipids form a bilayer (Fig. 3.11). One layer is oriented with its head groups on the upper layer of the membrane. The opposing layer on the opposite side has the opposite orientation such that the head groups point down. The resulting membrane has two polar surfaces and a central hydrophobic region composed of the hydrocarbon chains on the fatty acids. Not surprisingly, the polar heads are oriented such that they interact with water and the hydrophobic region is relatively free of water. As a result, the lipid bilayer forms a hydrophobic barrier around the cell through which water passes with great difficulty. The polar heads ionize in water and have close associations with water molecules. These polar and ionic interactions serve to stabilize the membrane in the configuration shown in Figures 3.11 and 3.12.

As depicted in Figures 3.11 and 3.12, the proteins floats about in the lipids. Many of the proteins actually extend through the membrane such that they

```
                          CH3
                          |
                   CH3 — N+ — CH3
                          |
                          CH2
                          |
                          CH2
                          |
                          O
                          |
                      O = P — O-
                          |
     H          H         O
     |          |         |
 H — C  ——————  C ——————  CH2
     |          |
     O          O
     |          |
     CH         C = O
     ||         |
     CH         CH2
     |          |
     CH2        CH2
     |          |
     CH2        CH2
     |          |
     CH2        CH2
     |          |
     CH2        CH2
     |          |
     CH2        CH2
     |          |
     CH2        CH2
     |          |
     CH2        CH
     |          ||
     CH2        CH
     |          |
     CH2        CH2
     |          |
     CH2        CH2
     |          |
     CH2        CH2
     |          |
     CH2        CH2
     |          |
     CH2        CH2
     |          |
     CH3        CH3
     |          |
     CH2        CH2
     |          |
     CH3        CH3
```

Fig. 3.13. The structure of a diglyceride molecule, in this case phosphatidylcholine.

are in contact with both the external fluid and the internal aqueous cytoplasmic fluid space. You will remember that amino acids with ionized side groups produce charged regions of the peptide chain. These portions of the protein tend to be associated with water molecules. Other amino acids have neutral side chains and they tend to be hydrophobic in nature. Proteins associated with membranes tend to assume a configuration in which the hydrophilic and the hydrophobic interactions are maximized, leading to the lowest energy configurations of the protein. As a result, highly charged regions of the protein will be found adjacent to the external and internal aqueous solutions, allowing the ionized amino acids to interact with water. The portions of the protein embedded in the internal regions of the lipid bilayer will possess an abundance of neutral amino acids with neutral side chains. The hydrophilic interactions of the proteins on either side of the membrane, and the hydrophobic interactions in the lipid bilayer tend to stabilize the proteins in the membrane. It is difficult for the protein to pop out of the membrane because this would bring amino acid side groups into an environment that is not energetically favored.

As we discussed above, proteins are linear arrays of amino acids. One might expect, therefore, that a protein would stick through the membrane and have two charged ends with a central hydrophobic region. When cell biologists began to study membrane proteins in more detail they found that most membrane proteins did not fit such a model. Instead most membrane proteins have multiple highly charged regions and multiple uncharged regions along their length. Second, the proteins were too long to extend only once across the membrane. It soon became clear that the proteins were arranged in the membrane not in a linear, stretched configuration, but rather in a highly folded cluster. In fact, a single membrane protein can extend back and forth across the membrane many times, a configuration referred to as a multipass protein (Fig. 3.14). Although the configuration of the protein in the membrane may seem unusual, it is yet another example of a protein spontaneously folding into a preferred configuration based on its amino acid sequence and dictates the surrounding environment. The protein is arranged in the membrane such that regions that are highly hydrophobic are embedded in the lipid bilayer and regions that are highly charged tend to be protruding on one surface or the other into the aqueous medium.

3.5.2 Integral and peripheral membrane proteins

When cell biologists began investigating membranes from cells, they discovered, not surprisingly, that these membranes consisted principally of lipids and proteins. They found that some proteins were hard to isolate unless one treated the membranes with detergent. This treatment dissolved and dissipated the lipid bilayer, and freed the proteins for further isolation and analysis. Some membrane proteins, however, could be isolated simply through the application of a strong salt solution (e.g., a concentrated KCl solution). It was found that the concentrated salt was competing for ionic bonds that held these proteins to the membrane, thereby releasing

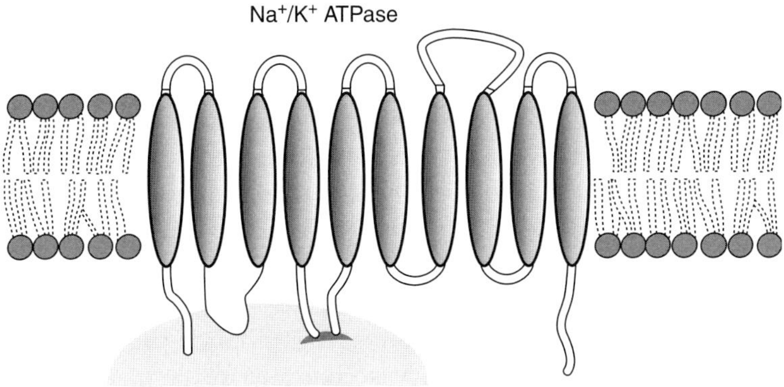

Fig. 3.14. The structure of multipass protein in a membrane.

these proteins from the membrane lattice. It became clear, therefore, that some proteins were securely embedded in the lipid bilayer while others were apparently more external and were held in place by ionic bonds. The former, embedded proteins were termed integral membrane proteins, while the nonembedded proteins were termed peripheral membrane proteins. Figure 3.12 illustrates our current understanding of the relative positions of these two classes of membrane proteins.

3.5.3 Proteins can act as pores in the lipid bilayer

If a multipass protein is large enough, the lateral surfaces of the protein adjacent to the lipids and within the bilayer can be hydrophobic while most of the internal, folded regions of the protein can possess ionized amino acids that hold the protein together and promote protein stability. This would result in a protein that spans the membrane and has strong hydrophobic anchors holding the protein in place, yet possesses a central hydrophilic core that can extend completely through the protein. This central hydrophilic core would form a channel running through the membrane.

Charged molecules such as ions or carbohydrates find the charged interior of such transmembrane proteins an appropriate environment for transient ionic binding. If an ionized channel were to extend completely across the bilayer, charged molecules that are unable to transverse the lipid bilayer could cross the membrane through such a charged core.

Cell physiologists have studied the properties of membranes by producing artificial lipid bilayers. These structures, referred to as black lipid membranes, consist entirely of membrane lipids structured in a lipid bilayer. They are highly impermeable to water ($p = 10^{-5}$ m/s) and to ions. Although the water permeability of a pure lipid bilayer is quite low, it was found that the addition of membrane proteins increased the osmotic permeability of the membrane. The proteins seem to provide sites where water can "leak" or diffuse through the membrane, perhaps along the sides of the proteins or through some central ionized core. Researchers found that if they added proteins to the lipid bilayer they could also increase the permeability of the membrane with regard to specific charged solutes. For example, adding the small peptide valinomycin increases the permeability of the membrane to K^+ but not to sodium or calcium. Adding the protein Glut1 increases the permeability to glucose but not to other sugars. Such proteins came to be called channels because they serve as conduits for charged molecules to cross the membrane. They do not require the specific input of energy to function as a channel and the molecules diffuse down their activity gradient through these molecules.

The specificity of the channels for one ion molecule, but not others, is apparently determined by the ionized sites in the central core of the channel. The charged side groups on the amino acids lining the central channel influence the binding and transverse passage of molecules based on their specific shape, size, and charge. These characteristics are different for each type of solute molecule, allowing the channels to impart highly specific permeability characteristics to the membranes.

Although all integral membrane proteins seem to increase the permeability of the membrane to water, in the 1980s proteins were discovered that took this permeability to new heights; they are selectively permeable to water. These proteins, termed aquaporins, are relatively impermeable to ions, but are highly permeable to water molecules. It seems that their physiological function is to increase the osmotic permeability of membranes. As we will see in subsequent chapters, this facilitates the process of moving water through epithelia and serves to make cells more rapidly responsive to osmotic gradients.

Over the course of evolution, some proteins that have specific ion-binding characteristics were functionally combined with proteins that can harvest chemical energy from high-energy organic bonds. These amalgamated proteins are capable not only for transferring ions across the membrane, but also for coupling such movements to energy provided from chemical reactions. As a result, these proteins can transport solutes against their electrical and chemical gradients. These proteins are termed ion transport proteins. Examples are the Na^+/K^+ ATPase, which transports these two ions across membranes in nerves and muscle, and the Ca^{2+} ATPase in skeletal muscle. These vital molecules will be described in more detail in Chapters 9 and 10. For the moment it is sufficient for us to recognize that these transport moieties consist of multipass membrane proteins, that they possess a central ionized core that provides the specificity required for their specialized function, and that their capacity to couple their function to energy sources allows them to create gradients for solutes across the membrane.

3.6 Membranes are ubiquitous

All animal cells possess membranes and these cellular structures play a central role in cell structure and function. The polar nature of water is critical in the formation of membranes as the phospholipids forming the bilayer cluster together with their polar heads pointed out. The hydrophobic nature of the lipids forms a tight and stable seal around the cell allowing cells to keep valuable compounds in and, equally vitally, many compounds out. Proteins in the membrane are influenced by their interactions with water. Their configuration, distribution, and orientation in the membrane are all dictated by hydrophobic and hydrophilic interactions. Finally, the interactions of integral membrane proteins with peripheral membrane proteins and indeed with other proteins in the cell are strongly influenced by pH, ionic strength, and the activity of water. We can see, therefore, that the regulation of the activity of water is critical for the proper configuration and activity of virtually every organic structure in the cell.

Suggested additional readings

Alberts, B., D. Bray, J. Lewis, M. Raff, K. Roberts & J.D. Watson (1989) *Molecular Biology of the Cell*. Garland Publishing, New York, NY.

Engelman, D.M. (2005) Membranes are more mosaic than fluid. *Nature* 438:578–580.

Paula, S. & D.W. Deamer (1999) Membrane permeability barriers to ionic and polar solutes. In *Membrane Permeability*, D.W. Deamer, A. Kleinzeller & D.M. Fambrough, eds. Current Topics in Membranes 48:77–95.

Singer, S.J. & G.L. Nicolson (1972) The fluid mosaic membrane model of the structure of cell membranes. *Science* 175:720–731.

Verkman, A.S. (1992) Water channels in cell membranes. *Annu. Rev. Physiol.* 54:97–108.

Yancey, P.H. (2005) Organic osmolytes as compatible, metabolic and counteracting cytoprotectants in high osmolarity and other stresses. *J. Exp. Biol.* 15:2819–2828.

4 The need for, and often conflicting issues of, osmoregulation and volume regulation

4.1 Osmotic regulation

The presence of solutes affects the activity of water. This activity in turn profoundly influences the structure and function of proteins. In addition, if differences in the activity of water exist across a semipermeable membrane (i.e., across virtually all biologic membranes), osmotically driven movement of water across the membrane will result.

It follows that animals can influence protein function and fluid movement by controlling the activity of water. A corollary of that statement is that it is critical for animals to regulate the activity of water if they are to maintain homeostasis. In particular, because the differences in the activity of water will drive fluid movement across membranes, the similarity or difference in the osmotic activity of fluids separated by a membrane is of critical importance.

For this reason, we now need to consider the activity of water in the various compartments that are separated by a membrane. Let us begin by considering a single cell residing in a fluid that is isosmotic with the cytoplasm of the cell (Fig. 4.1). If the fluid becomes slightly more concentrated, the activity of water in the cell would be higher than in the surrounding fluid and there would be a net outward diffusion of water from the cell. When this process reaches equilibrium, the cell will be isosmotic to the external fluid and the volume of the cell will be slightly less than before (Fig. 4.1). Alternatively, if the surrounding fluid shown in Figure 4.1 were to become slightly more dilute, water will show a net diffusion into the cell cytoplasm, causing the cytoplasm to become more dilute and the cell to swell. It can be seen, therefore, that changes in the osmotic concentration

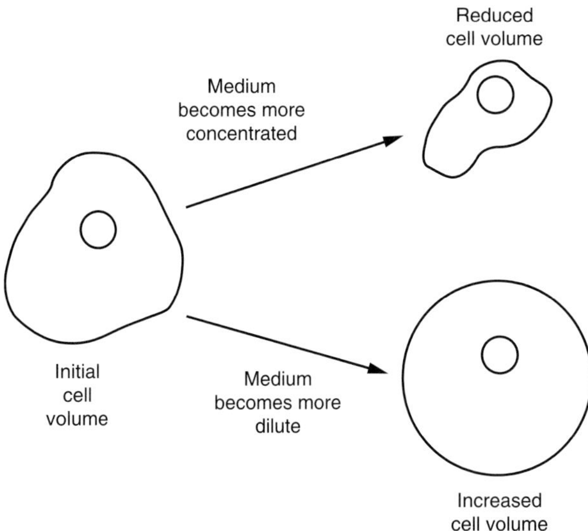

Fig. 4.1. The cell on the left has the normal, homeostatically regulated volume. If the medium becomes more concentrated, the cell will, in the short term, shrink due to osmotically driven loss of water from the cytoplasm. If the medium becomes more dilute, the cell will, in the short term, swell due to osmotically driven entry of water into the cytoplasm.

of the surrounding fluid can have two disparate effects on cells. They can affect both the osmotic concentration and the volume of the intracellular compartment. These two parameters are linked but distinct.

For single-celled organisms such as protozoa, we need to consider only these two fluid compartments: the extracellular fluid compartment and the intracellular cytoplasmic compartment. Marine protozoa generally are isosmotic with the external oceanic milieu. Osmotic regulation for them is achieved by the unchanging nature of the vast oceans (see Chapter 5). Freshwater protozoa have cytoplasmic osmotic concentrations that are much higher than the external medium. This causes them to gain water constantly through osmosis and they have mechanisms for excreting the excess water they gain (see Chapter 7).

Over evolutionary time, animals evolved from a single cell state to more complex, multicellular body plans. Very early in evolution, animals arose whose body plan involved the cells being bathed by an internal fluid. In other words, animals evolved an integument which faced the external environment, allowing the internal organs to be bathed by an interstitial fluid between the cells. This, on the one hand, served to protect and partially isolate the cells from the external environment, but it also meant that transport mechanisms involved in osmotic regulation had to be located at this external barrier. The body plans of animals evolved in a number of different directions. In animals such as flatworms, the external surface was derived from ectoderm during development while the gut was derived from endoderm (Fig. 4.2). The spaces between the integument and gut filled with cells derived from mesoderm. In pseudocoelomate animals such as nematodes, a fluid-filled space known as the pseudocoelom forms around the gut during development (Fig. 4.3). In coelomates, both the ectoderm and the endoderm are lined with mesoderm. A fluid-filled space termed coelom surrounds the gut which is also lined with mesoderm (Fig. 4.4). In animals, such as vertebrates, that have a closed circulatory system, another fluid compartment, the blood, exists that is distinct from the interstitial and coelomic fluids. Despite these complex and highly variable body plans, the osmotic consequences of multicellular body plans are rather simple. In all multicellular animals possessing an integument and various types of internal compartments, the intracellular compartment is always isosmotic to the extracellular fluid in which it is bathed. From an osmotic point of view, therefore, it is apparent that in order to protect and regulate cellular function, regulation of the extracellular fluids of multicellular animals is critical.

Let us examine the consequences of these facts by using the example of a marine flatworm. Even in animals as morphologically simple as a flatworm, the cells of the body are bathed by body fluids which reside inside the integument. Imagine now that the flatworm finds itself in a tide pool into which rain falls for a short period of time, diluting slightly the water in which the flatworm resides. As the dilute rain mixes into the tide pool, the activity of water in the pool will rise and water will move osmotically

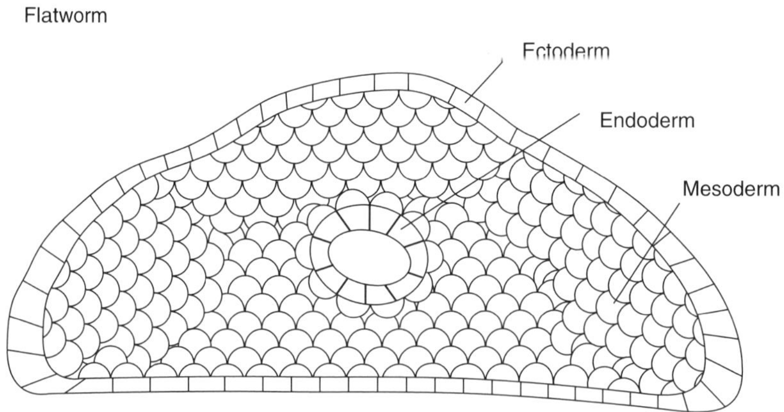

Fig. 4.2. A cross section of a flatworm showing the organization of internal compartments and cell layers.

across the integument of the flatworm and into the flatworm's body fluids. It is important to realize that this change in osmotic concentration will extend throughout the body of the animal and into all of the tissues in the body. Water will diffuse down its activity gradient until all of the osmotic

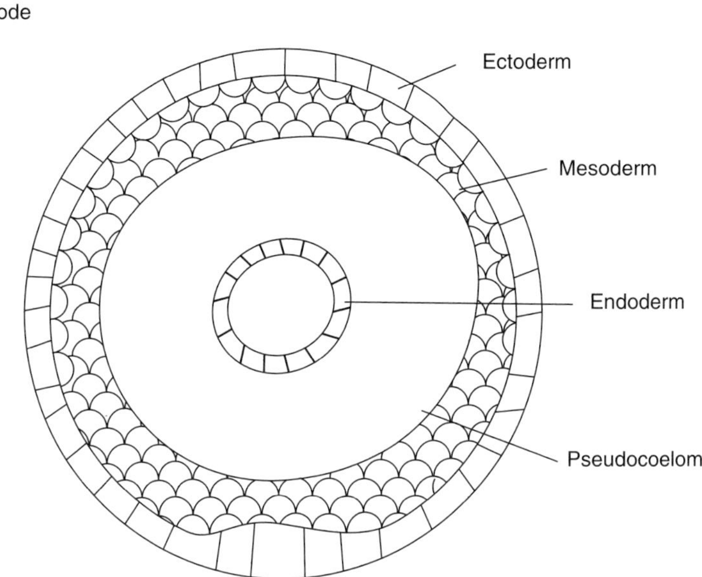

Fig. 4.3. A cross section of a nematode showing the organization of internal compartments and cell layers.

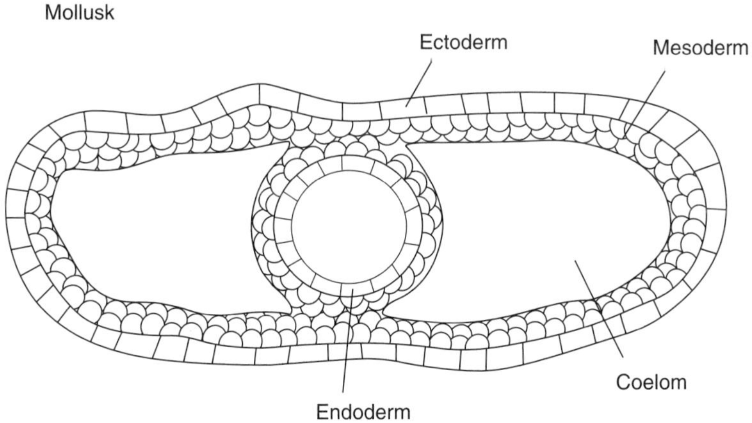

Fig. 4.4. A cross section of a mollusk showing the organization of internal compartments and cell layers.

gradients across membranes come into equilibrium. An additional effect will be that the flatworm's body volume will increase.

Let us consider a different animal with different physiological capacities, namely, a goldfish living permanently in dilute water. As we will discuss further in Chapter 8, the fish maintains an osmotic concentration in its body fluids that is well above that of dilute, freshwater. As a result, water moves osmotically across the integument and into the blood of the fish. All of the internal fluid compartments, whether they be blood, interstitial fluid, coelomic fluid, or cytoplasm, are isosmotic with one another. Therefore, if the fish allows its blood to become dilute, then this would affect every cell in its body. Rather than allowing this to happen the fish uses its kidneys to produce dilute urine, thereby reducing the osmotic concentration of the blood (see Chapter 8 for more details). By regulating the osmotic concentration of the blood, the kidneys regulate the osmotic concentration of all of the fluid compartments in the body. The internal environment is entirely at the mercy of the osmotic concentration of the extracellular fluids.

4.2 Volume regulation

We have seen in the above examples that the osmotic entry of water into a fluid compartment alters the osmotic concentration of that compartment. It has the added effect of changing the volume of that compartment. Similarly, water moving out of a compartment reduces the volume of that compartment. We normally do not worry about the volume effects of ions moving, except in the context of the effect of those ions on further water movements. The reason for this can be appreciated if we consider the makeup of physiologically relevant solutions.

As we will see in subsequent chapters, the internal body fluids of terrestrial and freshwater animals have an osmotic concentration of around 300 mOsm. This is equivalent to the osmotic activity of 0.3 moles of osmotically active solute per kilogram of water. The molecular weight of water is 18 Da and the weight of 1 mole of water is therefore 18 g. As shown Eqn 4.1, 1 kg of water therefore contains 55.56 moles of water.

$$\frac{1000 \text{ g of water}}{18 \text{ g of water/mole}} = 55.56 \text{ moles} \tag{4.1}$$

Using this information, we can see that in 1 liter of a solution with an osmotic concentration of 300 mOsm, there are 55.56 moles of water and 0.3 moles of solute. Therefore, there are 185.2 water molecules for every solute molecule present.

$$55.56 \text{ moles of water}/0.3 \text{ moles of solute} = 185.2 \text{ moles/mole} \qquad (4.2)$$

The movement of one solute molecule across a membrane that is at osmotic equilibrium in the body of these animals would cause the movement of 185 water molecules to restore osmotic equilibrium. The movement of ions, therefore, has a direct effect on volume regulation and fluid transport, but only because it so profoundly affects the distribution of water.

Changes in osmotic concentration and volume can have both deleterious physiological effects, yet the two problems often require slightly different mechanisms to correct. Let us examine a specific example to understand this point. Consider osmotic and volume regulation in a mammal. If the mammal suffers a serious injury, blood is lost until such time as the blood clots and the wound can be sealed to avoid further fluid loss. At that point, the osmotic concentration of the extracellular fluids would be largely unchanged but the volume of the extracellular fluids and, particularly, the blood would be reduced below the normal physiologically regulated value. The challenge for this animal is to restore the lost blood volume while maintaining appropriate osmotic regulation. As we described above, the volume in physiological solutions is largely water. The animal could replace this volume by drinking water. As this water moves across the gut and into the blood, however, it would upset osmotic homeostasis by diluting the blood. This added water would have to be removed by the kidneys to restore the proper blood osmotic concentration. The result would be no change in blood volume.

Alternatively, the animal might seek to replace the blood volume by ingesting salt (NaCl). As this salt moves across the gut, it would raise the osmotic concentration of the blood, drawing water from the intracellular compartment. This would transiently raise blood volume, but again, homeostatic osmotic controls would induce the kidneys to remove salts through the production of concentrated urine. As the urine can never be made as concentrated as a pure salt, the ingestion of salt would actually lead to a net reduction of blood volume by the time the proper osmotic concentration has been restored. We can see that in these examples, osmotic and volume regulation can work at cross purposes.

To replace the blood volume lost, the mammal must obtain both salts and water in order to produce a fluid of the appropriate osmotic composition. This requires both the acquisition of the appropriate compounds and their appropriate regulation by the excretory organs. I mention this to make the point clear that both osmotic and volume regulations are vital issues for animals. The challenge from the physiological point of view is that the solutions to one of these regulatory challenges may at times interfere with the regulation of the other.

Table 4.1. Examples of major compatible organic osmolytes in animal groups

Mollusks	Betaine, taurine, glycine
Elasmobranchs	Urea, trimethylamine oxide
Annelids	Sorbitol, betaine, taurine, glycine,
Insects	Proline, trehalose

4.3 Compatible solutes

By competing for hydrogen bonds that hold the protein in its active configuration, highly charged compounds can inactivate (denature) enzymes and structural proteins. There are classes of solutes, however, that interact with proteins in a manner such that they stabilize proteins in the active configuration. Such solutes are termed "compatible solutes." Table 4.1 is a list of compatible solutes and the organisms from which these molecules have been identified and isolated. It can be seen that amino acids lead the list. In addition, a number of sugars and sugar alcohols can serve as compatible solutes.

Compatible solutes serve two purposes in the cells of euryhaline animals. One purpose is that their presence helps to promote and protect the appropriate configuration of proteins. They serve this purpose even if the proteins are subjected to slight stresses due to changes in temperature, pH, or salinity. The second purpose served by compatible solutes is that they can be varied in concentration with little effect on protein structure. In those animals that experience substantial changes in cellular osmotic concentration (see osmoconformers in Chapter 5), the cells can vary the concentration of compatible solutes as a means of regulating cell volume. Although changes in noncompatible solutes would affect protein function, changes in compatible solute concentration is relatively benign in its effects on protein structure and function. When we examine the cells of euryhaline osmoconformers, therefore, we find that their cells are rich in compatible solutes. The concentration of these solutes can be modified up or down to regulate cell volume while simultaneously retaining appropriate protein structure and function.

Suggested additional readings

Gilles, R. (1979) *Mechanisms of Osmoregulation in Animals: Maintenance of Cell Volume.* John Wiley & Sons, Chichester, NY.

Johnson, L.G. (1987) *Biology.* Wm. C. Brown, Dubuque, IN.

Yancey, P.H. (2005) Organic osmolytes as compatible, metabolic, and counteracting cytoprotectants in high osmolarity and other stresses. *J. Exp. Biol.* 15:2819–2828.

5 Osmoconformers

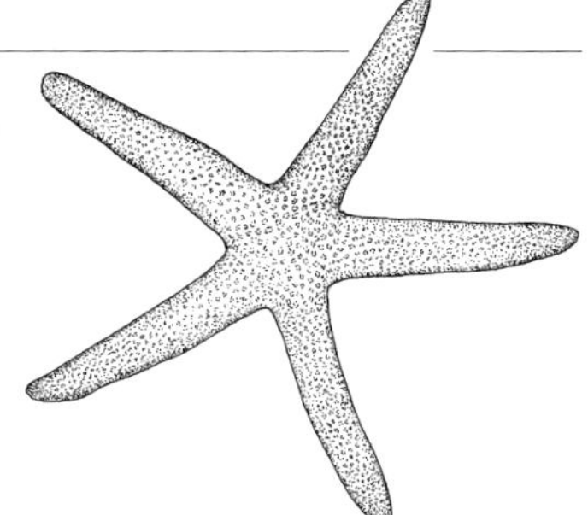

5.1 Introduction

In the marine environment, the vast majority of organisms are osmoconformers. That means that their extracellular fluid conforms to (is isosmotic with) the surrounding seawater and no mechanism exists for maintaining osmotic gradients relative to that medium. As discussed in Chapter 4, because the cells of all organisms are also isosmotic with the extracellular fluid surrounding them, this means that in osmoconformers, the entire organism is subject to the vagaries of the external environment.

It behooves us, therefore, to determine just how variable the marine environment is, and to learn a bit more about the osmotic and ionic concentrations of seawater. The open ocean is actually surprisingly constant in terms of its osmotic and ionic makeup. The open Pacific Ocean has an average salinity of 35 ppt. This means that if you took 1000 g (close to 1 liter) of seawater from the open Pacific Ocean and evaporated all of the water from that, you would be left with 35 g of salts. Table 5.1 shows the concentration of other regions in oceans. It can be seen that the open oceans are well mixed at their surfaces. A few areas of the oceans that are somewhat more isolated from mixing currents can vary from the norm. This includes enclosed regions in warm, dry climates such as the Red Sea, and enclosed basins in cold climates with multiple inflowing rivers, such as the Baltic Sea.

Although the total osmotic concentration of the oceans can vary, the ionic ratios in these various bodies do not vary at all. In other words, although the Baltic Ocean may have an osmotic concentration one-half that of the open Atlantic Ocean, the ratios of sodium to chloride will be identical in both oceans. The regions in the ocean that are more concentrated, or less

Table 5.1. The salinity of ocean waters

Geographic region	*Salinity (ppt)*
Open Pacific Ocean	33–37
Open Atlantic Ocean (20° North)	36
Open Atlantic Ocean (60° North)	31
Arabian Sea (relatively open body in an arid region)	37
Red Sea (relatively isolated body in an arid region)	40
Pacific Ocean (near Alaskan coast)	24–32
Puget Sound (relatively open body near the mouth of large rivers)	21–27
Baltic Sea (relatively isolated body in a cold, well-watered region)	5–15

so, vary only by the amount of water diluting the ions. The ratios of all the constituents remain constant. As a result, marine organisms may have to adjust to slight variations in the total osmotic concentration of the external medium, but the ions present and the ratios of these ions are totally invariant in the marine habitats.

5.2 Marine invertebrates

5.2.1 Echinoderms

A useful place to start our discussion on osmoconformers is with the Echinoderms, for example, the sea stars. Sea stars are restricted to stable, marine environments. These animals not only have an extracellular fluid which is isosmotic to seawater, but also their extracellular fluid is seawater. Echinoderms have a sieve-like plate (termed madreporite) on their external surface which serves as a filter. Fluid can flow in and out between the extracellular compartment and the seawater through this plate. As a result, the internal fluid of these animals is identical to seawater in every aspect, with the exception that large particles and organisms, such as bacteria, protozoa, and larvae, are excluded. The cells are, therefore, bathed in a NaCl-rich fluid, which varies very little from day to day and indeed from century to century as long as the sea stars are bathed by water from the open ocean. Sea stars expend no energy on the regulation of the ionic and osmotic makeup of the extracellular fluids as these properties are maintained by the mixing properties and stability of the vast oceans.

Seawater is ideal as an extracellular fluid. It is rich in sodium and relatively low in potassium. Such an ionic mix is required in the extracellular fluid for the proper functioning of action potentials in nerves and muscles. Seawater is highly stable in osmotic concentration so the cell volume is not constantly challenged by osmotic shifts. It contains a mix of minerals

Table 5.2. Elements in seawater (data from Weast and Astle, 1980)

Element	*Concentration in parts per million*
Chloride	18,980
Sodium	10,561
Magnesium	1,272
Sulfur	884
Calcium	400
Potassium	380
Bromide	65
Aluminum	2
Fluoride	1.4
Nitrogen	0.05
Phosphorus	0.01
Iodine	0.05
Iron	0.02
Zinc	0.005
Copper	0.09
Manganese	0.01

necessary for proper cellular function such as calcium, magnesium, iodine, and phosphorus. Table 5.2 shows the extraordinary diversity of vital minerals and nutrients contained in seawater.

Most scientists believe that life first evolved in the oceans (an alternative theory evokes hot springs). Seawater was unquestionably, however, the external medium in which the early eukaryotes evolved. It is a tautology, therefore, to argue that seawater is a good external milieu for living organisms. The eukaryotes evolved with seawater as the external medium and natural selection weeded out organisms not suited to such an environment. As a result, animals possess cellular processes that function well and indeed generally require a sodium-rich extracellular fluid resembling seawater with regard to a number of key parameters.

Seawater, however, is totally inappropriate as an intracellular fluid. Proper functioning of the action potential requires a strong gradient for sodium across the plasma membrane of excitable cells. If seawater were present as both the extracellular and the intracellular fluid, then nerves and muscles could not function. In addition, most intracellular proteins are denatured by the high sodium concentration of seawater. Instead, intracellular proteins require high levels of potassium and low levels of sodium and calcium. For these reasons, the intracellular milieu in animals, including osmoconformers, is ionically very distinct from seawater (see Table 5.3). The cells must

Table 5.3. The principal ions in seawater (data from Potts and Parry, 1964) and in the muscle of the squid *Sepia officinalis* (data from Potts and Parry, 1964; Robertson, 1965)

Ion	*Concentration in seawater (mmol/kg)*	*Concentration in the cytoplasm (mmol/kg)*
Sodium	478	31
Magnesium	55	19
Calcium	11	2
Potassium	10	189
Chloride	558	45
Sulfate	28	2
Bicarbonate	2	Variable

Some of the elements in the cytoplasm may not be fully ionized.

expend substantial amounts of energy constantly pumping ions into and out of the cells to maintain the necessary elemental distinction between extracellular fluid and intracellular fluid (see Chapter 9 for more detail on ion transport).

While I am emphasizing here the fact that the intracellular fluid is ionically quite distinct from the extracellular one, I also wish to remind the reader that the osmotic concentration of the cytoplasm is identical to that of the extracellular fluid. Therefore, the cell cytoplasm will differ substantially in ionic makeup from seawater, but the total osmotic concentrations of the two compartments remain always equal.

It is not sufficient for the sea star to regulate the intracellular milieu by substituting potassium and bicarbonate for sodium and chloride. The cell interior is rich in organic compounds required by the cells for proper function. These include sugars, amino acids, and nucleotides. These are all small, highly soluble compounds with high osmotic activity. In addition, the cell interior contains a myriad of proteins in the form of enzymes, cell regulators, structural proteins, and cofactors. These also contribute substantially to the osmotic concentration of the intracellular fluid. Because (1) the intracellular fluid is rich in organic compounds, (2) the extracellular fluid is low in organic compounds, and (3) the two compartments must be equal in total osmotic concentration, it follows that the total ionic concentration of the intracellular fluid is lower than that of the extracellular fluid. Although the activity of water in the intracellular and extracellular compartments is equal, the intracellular compartment differs in that the ions present are different, the total ionic strength is lower, and the contribution of organic compounds to the total solute concentration is much greater (see Table 5.3).

The intracellular fluid in sea stars is rich in organic molecules. In addition to the proteins and carbohydrates required for intermediary metabolism, the cells of sea stars accumulate relatively high concentrations of amino acids. These small organic molecules occur in low concentrations in seawater and many marine organisms can accumulate them through active uptake (see Chapter 9). In addition, ingested proteins can be broken down into their constituent amino acids providing an abundant source of small osmotically active molecules. For example, an average mid-sized protein might have a molecular weight of 100,000 Da. In fact, this protein would have the osmotic activity of a single molecule. If it were broken down into its constituent amino acids, then the osmotic activity associated with this material would increase substantially. Amino acids have an average molecular weight of about 100 Da, so splitting the protein into single amino acids would provide 1000 small molecules. As a result, the osmotic activity associated with this organic material would increase 1000-fold. It is clear, therefore, that cells can decrease their cytoplasmic osmotic activity substantially through protein synthesis, thereby reducing free amino acids, or increase osmotic concentration through protein degradation, which serves to increase free amino acid levels.

In sea stars, the extracellular fluid is filtered seawater and, as a result, they are totally dependent on the external medium for osmotic regulation. Nonetheless, the intracellular fluid differs substantially from seawater in its ionic makeup. The intracellular fluid is rich in potassium, bicarbonate, and magnesium while the extracellular fluid is high in sodium and chloride. In addition, abundant organic molecules contribute to the internal intracellular osmotic concentration, including proteins, amino acids, and organic acids. Sea stars are considered to be stenohaline. Their cells will die if the total osmotic concentration varies very much from that of the open ocean. If you go to tide pools along the ocean shore, you will find sea stars exposed to the air only during the lowest tide. They do not venture far up into the high intertidal where they might, on the one hand, be exposed to dry air for extended periods, or equally deadly, to diluting rain. If a sea star does become dislodged by the waves to be swept up on shore, then it very often dies due to the inability to osmoregulate. Sea stars and many other marine organisms are tied to and dependent on the stabile, nurturing constancy of the oceans.

5.2.2 Mussels

Marine mollusks breathe by means of gills. In order to facilitate rapid gas exchange, gills must be thin-walled and filamentous. The gills of mollusks are covered with thin epithelial cells and the blood is transported into each gill filament in branchial blood vessels (Figs. 5.1 and 5.2). The shell of the mussel is covered by a thin delicate mantel as well. Mollusks, therefore,

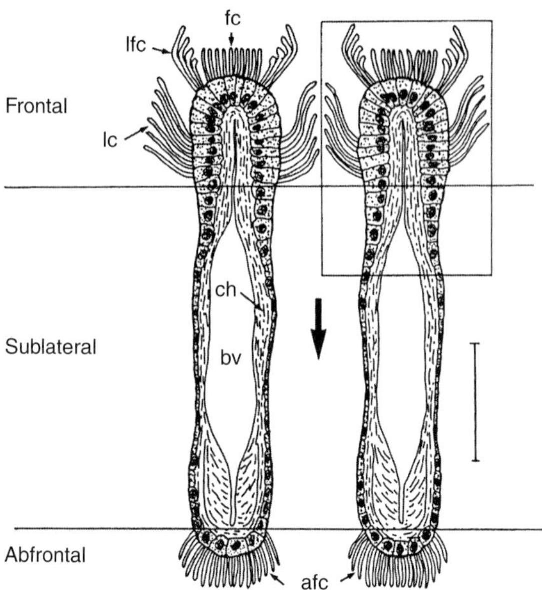

Fig. 5.1. A diagrammatic representation of a cross section of the gill of a marine mollusk. The figure-like extensions on the frontal and abfrontal cells are cilia that beat and move water over the gill. afc, abfrontal cilia; bv, blood vessel; ch, conchiolin, a compound that stiffens the gill; fc, frontal cilia; lfc, laterofrontal cilia; lc, lateral cilia. Scale bar = 50 µm. (Adapted from Wright *et al.*, 1987.)

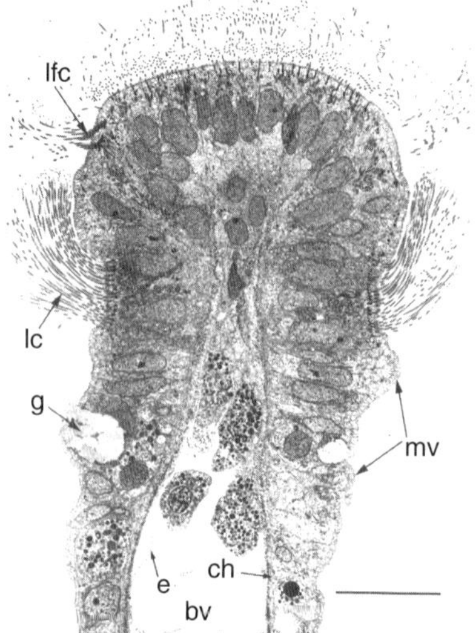

Fig. 5.2. An electron micrograph showing a cross section of the gill of *Mytilus californianus*. The view is identical to that shown within the box in Fig. 5.1. bv, blood vessel; ch, conchiolin; e, endothelial cell; g, goblet cell; lfc, laterofrontal cilia; lc, lateral cilia; mv, microvilli. Scale bar = 20 µm. (Adapted from Wright *et al.*, 1987.)

have a large surface to volume ratio and delicate, thin epithelia in intimate contact with their aqueous environment. As a result, marine mollusks are by necessity osmoconformers.

In contrast to sea stars, which are stenohaline, mussels are euryhaline, by which we mean that they can survive in a broad range of salinities. If you go to harbors in the mouth of rivers, you will find abundant clusters of mussels coating the piers, rocks, and shores. The mussels can survive extended exposure to air during low tide. They also rapidly resume feeding when the tide comes back in. Finally, and most critically, they can also survive and feed if heavy rains swell the river filing the harbor with more dilute water. How are mussels able to survive in these variable environments given their requirement to osmoconform?

In part, mussels survive in diverse osmotic environments by closing their shells tightly during times of adverse environmental conditions. This is quite evident when you buy mussels at the fish market. Every living mussel is tightly shut, awaiting better environmental conditions. It is important that mussels can seal themselves off from dry environments as mussels have little capacity for maintaining an osmotic gradient between the external environment and their extracellular fluid.

In addition to simply isolating themselves from adverse conditions, mussels possess biochemical and physiological mechanisms that allow the cells throughout the body to survive while covarying with the changing external osmotic concentrations. In seawater, the cells of mussels are similar to those of sea stars. They contain high concentrations of potassium, proteins, and amino acids. A substantial portion of these solutes are compatible osmolytes that serve to protect the structural integrity of the proteins at high osmotic concentrations. Unlike the sea stars, however, mussels can survive in salinities as high as seawater and as low as one-third strength seawater. Let us examine the processes that allow such a broad range of salinity tolerance.

In full strength seawater, the cells of the mussels have an osmotic concentration equal to that of seawater, about 1000 mOsm. This high osmotic concentration in the cells is achieved by the accumulation of potassium, protein, organic acids, and amino acids. Now consider the consequences for these mussels, living in seawater, when the tide swings toward low tide and water begins to recede from the mouth of the river. Freshwater moves down from the river to partially replace the seawater. As a result, the salinity around the mussel begins to drop. Under these circumstances the activity of water in the external medium is increasing. Water will, therefore, flow by osmosis into the mussel's blood and into the cells. This would produce a crisis for the mussel involving both cell volume regulation and cell osmotic regulation. To regulate cell volume, the mussel cells must reduce the number of solute molecules present in the cytoplasm. This is achieved in two ways. The first mechanism is to transport solutes out. This is achieved in the form of outward transport of potassium and amino acids. In response to the cell swelling, volume-sensitive channels open in the plasma membranes of the cells, allowing potassium and amino acids to leak down their concentration gradients (see Chapter 11). Physiologists have measured immediate increases in the concentrations of amino acids (particularly aspartate and taurine) circulating in the blood of mussels following down transfer to more dilute waters. These solutes derive from the cells. The cells must "dump" these solutes as a way of regulating cell volume. If they did not, the cells would rapidly swell and burst. A second mechanism for reducing the quantity of internal solutes consists of the synthesis of proteins from free amino acids in the cell. By combining individual free amino acids into a single large protein molecule, the number of osmotically active molecules in the cell is greatly reduced.

Cell volume regulation in mussels can therefore be achieved by tying the transport of solutes and the synthesis of proteins to cell volume measurements. The precise mechanisms by which this is controlled are still under investigation. Suffice it to say that the control of solute retention in each cell is influenced by stretch receptors in the cell membranes that provide the cell with information about the degree of tension on the cell membranes occurring due to changes in cell volume.

To understand more fully the challenges that cells face with changes in external salinity, let us consider the following scenario. Imagine a mussel in full strength seawater suddenly being faced with a rapid transition to one-half strength seawater. This would be a substantial change, but one faced by many estuarine animals. If no solute regulation occurred, water would flow by osmosis into the cell until the cell reaches twice its original volume. At that point, the internal and external osmotic concentrations would be identical and net water flow would cease. Most cells cannot survive this level of volume increase. It is clear, therefore, that the task facing the cell is not simply to adjust the osmotic concentration of its cytoplasm, but rather to adjust the total number of osmotically active solute molecules in the cytoplasm. In other words, to maintain homeostasis with regard to cell volume, the cell must reduce the number of osmotically active solute molecules. If the cell carries out these tasks and reduces the number of active solute molecules to one-half the previous number, the cell volume will return to the same volume it was previously. The osmotic concentration of the cytoplasm will be reduced to one-half its previous value. It can be seen, therefore, that the cells of estuarine mollusks are not regulating cytoplasmic osmotic concentration, in fact that varies enormously in concert with the environment. Instead, they are regulating cell volume very closely, and they do this by regulating the number of active solute molecule in the cell.

If we return to the problem of adjusting to a medium of one-half strength seawater, the cytoplasm of the cells contains abundant carbohydrates and organic acids necessary for cellular function and metabolism. The solutes that the cell gets rid of, either through macromolecular synthesis or leakage from the cytoplasm, must therefore be "extraneous" molecules not directly necessary for cellular function. It follows that in higher salinities, the cells of mussels contain abundant molecules that are present largely for reasons of volume regulation. These "extraneous" solutes are largely compatible solutes as defined in Chapter 4. Upon, dilution of the medium, these solutes can be rapidly jettisoned to restore cell volume and preserve cellular integrity.

5.2.3 Marine crabs

Marine crabs also respire by means of gills. As the gills possess thin, permeable epithelia that must remain in contact with the external aqueous medium, most marine crabs are osmoconformers. The hemolymph of these crabs has the same osmotic concentration as the external medium but the ionic makeup can sometimes differ considerably. In particular, the magnesium and sulfate concentrations in the hemolymph are often lower than that of seawater. In all species examined to date, the intracellular milieu of crabs contains high concentrations of compatible organic solutes. We have already discussed the important role that these solutes play in protecting metabolic function in the cytoplasm under both high and variable salinity conditions.

Some crabs are stenohaline, meaning that they cannot withstand substantial reductions in salinity. Other species are euryhaline osmoconformers, meaning that they osmoconform, but can withstand substantial changes in the external salinity. *Cancer gracilis* is an example of such a species. This species can survive in estuarine environments where the salinity of the surrounding water is somewhat depressed by freshwater flowing off the land. *Cancer* cannot survive indefinitely but, for example, at concentrations around 50% seawater the crabs can survive on average about 30 hr. This may not seem adequate for surviving in such environments as the whole population might be extirpated after a few days at this salinity. Remember, however, that the estuarine environment is subjected to twice-daily high and low tides. The crabs only have to be able to survive for 5–6 hours at low salinity before the next high tide would bring increasingly saline water back into their environment. By linking organic solute and potassium efflux to stretch receptors in the cell membranes crabs can, similar to other marine invertebrates, allow the osmotic concentration of the body fluids to reflect the changing external environment while constantly regulating and adjusting cell volume and blood volume.

5.2.4 Insects

A few species of insects can survive in saline water using osmoconformation as their osmotic strategy. These include a species of caddis fly and a number of mosquito species. Mosquito larvae in the genera *Culex* and *Culiseta* osmoconform in all media that are more concentrated than 30% seawater. In media ranging from about 300 mOsm to 1000 mOsm, the osmotic concentration of the hemolymph exactly matches that of the external medium. In media more dilute than 300 mOsm, the larvae osmoregulate in a manner identical to freshwater species.

When larvae of osmoconforming mosquito species are reared in concentrated media, the hemolymph contains high levels of amino acids, particularly proline and serine. The hemolymph also contained unusually high levels of the disaccharide sugar trehalose. The larvae are able to synthesize these compatible solutes in their cells and accumulate them in the hemolymph. In this manner, they raise the osmotic concentration of the hemolymph to a level equal to that of the external medium.

The cells of these osmoconforming larvae have a cytoplasmic osmotic concentration equal to the hemolymph and thus to the medium. Physiologists were interested in determining the intracellular osmolytes employed by these species. They found that trehalose, the common blood sugar of insects, was not found in the cells in measurable amounts. This sugar, which is an excellent compatible solute, is apparently strictly extracellular. Proline, however, was found in high concentrations inside the cells as well as in the hemolymph. This is an interesting example of a compatible osmolyte being used both intracellularly and extracellularly in an animal.

There is an important difference between the osmotic strategy of osmoconforming insects and those of osmoconforming marine invertebrates such as mussels and crabs. The marine invertebrates allow their blood or hemolymph to come into ionic equilibrium with the seawater, regardless of its strength. As a result, although the fluid surrounding the cell is variable in osmotic concentration it is reliably always rich in sodium and chloride. Insects are frequently found in inland or coastal saline waters in which the ionic ratios can be quite variable due to freshwater runoff or mineral input from the soil. These insects have evolved a strategy that involves allowing the hemolymph to osmoconform, but the osmolytes are organic solutes produced by the insect itself. This strategy allows the insects to take advantage of the energy savings inherent in osmoconformation without permitting the external ions entry into the body. Similar to marine invertebrates, the intracellular compartment of osmoconforming insects contains amino acids, as these have neutral or even beneficial effects on protein and membrane structure.

5.3 Osmoconforming chordates

Most chordates are osmoregulators, meaning that they expend metabolic energy to retain body fluids that differ substantially from the external environment and are stable. A few chordates are osmoconformers. We will review them here and examine the physiological processes employed by these animals.

5.3.1 Hagfish

Hagfish are osmoconformers. They are found exclusively in marine waters and their blood, and therefore their tissues as well, is isosmotic with the medium. The ionic makeup of their blood differs a bit from seawater, however, being lower in the concentrations of the divalent ions Ca^{2+}, Mg^{2+}, and SO_4^{2-}. Their blood is not particularly rich in organic compounds. It follows that the monovalent ions, particularly Na^+ and Cl^-, are somewhat elevated compared with the external seawater.

The advantages of osmoconformation are largely energetic. Because they are isosmotic to the external environment, hagfish do not experience large fluxes of water across the body wall. Compared to the vertebrates described in subsequent chapters, they expend little energy on ion transport or the movement of large volumes of water into the urine. The disadvantages of osmoconformation are that the tissues are subjected to any variabilities in osmotic concentration that occur in the external medium. As hagfish occur strictly in the open ocean, this poses little danger to them.

The cytoplasm of hagfish is enriched in potassium and organic solutes, including amino acids, in a manner similar to that of the invertebrates described above. They are stenohaline and have little capacity for adapting to dilutions of the medium.

5.3.2 Sharks and skates

Marine sharks and skates osmoconform. Their blood is isosmotic with the surroundings. As a result, the osmotic movement of water across the integument, including the gills, of shark and skates is minimal. Sharks and skates, therefore, do not drink the medium to maintain body volume. The flux of ions is also small and is restricted to the ions that diffuse across the gills or are contained in the seawater ingested with the meals.

Although the blood of sharks and skates is isosmotic to the surrounding seawater, the ionic concentrations of the blood differ substantially from those in the external medium. Whereas the sodium and chloride concentrations in seawater are 460 and 540 mM, respectively, the concentrations in marine sharks and skates are around 250 mM for each ion. It is clear that other solutes must be present to produce the remaining osmotic activity. The blood of sharks and skates contains high levels of urea (see Fig. 5.3), a nitrogenous waste produced during the catabolism of proteins. The concentration of urea can be as high as 370 mM. Being a highly soluble compound, urea contributes substantially to the osmotic activity of the blood. Urea is, therefore, an ideal solute for raising the osmotic concentration of the blood as it is cheap to produce (being a waste by-product) and of low molecular weight. The problem is that urea is a strong denaturing agent for proteins. In this sense, urea cannot really be considered a compatible solute, as it does not protect protein function. The solution to this physiological conundrum is the presence of a second organic compound, trimethylamine N-oxide (TMO), in blood of sharks and skates (Fig. 5.4). This compound serves to counter the effects of urea. The concentrations of urea alone would be toxic to sharks and skates were it not for the simultaneous presence of TMO.

Urea and TMO together represent a substantial portion of the osmotic activity both intracellularly and extracellularly in sharks and skates. The presence of these two organic compounds varies with the salinity of the external medium, allowing the ionic makeup of the blood and cytoplasmic compartment to be well regulated. As urea by itself is toxic, the term compatible solute seems inappropriate for this osmolyte. Yancey and Somero (1979) proposed the term counteracting solutes for pairs of solutes which together are benign but where one of the solutes alone would have deleterious effect on protein and membrane structure.

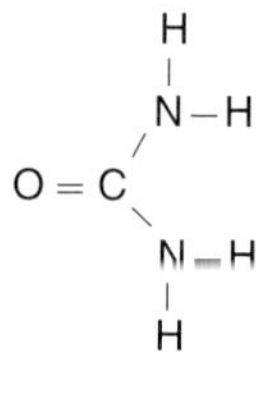

Fig. 5.3. The structure of urea.

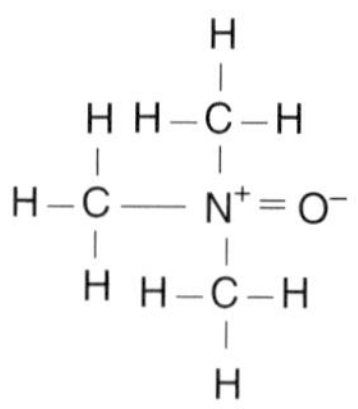

Fig. 5.4. The structure of trimethylamine N-oxide.

5.3.3 Crab-eating frog

One last vertebrate deserves mention as an osmoconformer. While most amphibians are restricted to freshwater and are hyper-regulators (see Chapter 7), one species of frog can be found in coastal marine habitats. This species is the crab-eating frog, *Rana cancrivora*. This frog cannot survive indefinitely in full strength seawater, but it is able to exploit estuarine environments with salinities well above those that would kill other amphibians. The frog forages into estuarine and intertidal waters, feeding on the invertebrates that are abundant there, including the crabs that give it its name. Frogs maintained in brackish waters have ion concentrations in their blood that are normal for an amphibian (Na^+ concentration = 250 mM, Cl^- concentration = 230 mM). The blood, however, is isosmotic with the medium with urea contributing a substantial portion of the additional osmotic activity. As the frogs move through the environment, experiencing changing osmotic conditions, water moves into and out of the body across the permeable skin in response to the osmotic gradient. The frogs are able to achieve volume regulation and maintain stable ionic concentrations in the body fluids by regulating the rate at which urea is either retained in, or excreted from, the body.

Suggested additional readings

Curtis, D.L., E.K. Jenson & I.J. McGaw (2007) Behavioral influences on the physiological responses of *Cancer gracilis*, the graceful crab, during hyposaline exposure. *Biol. Bull.* 212:222–231.

Patrick, M.L. & T.J. Bradley (2000) The physiology of salinity tolerance in larvae of two species of *Culex* mosquitoes: the role of compatible solutes. *J. exp. Biol.* 203(4):821–830.

Randall, D., W. Burggren & K. French (1997) *Eckert Animal Physiology*. W.H. Freeman and Co., New York.

Somero, G.N. (1986) From dogfish to dog: trimethylamines protect proteins from urea. *NIPS* 1:9–12.

Weast, R.C. & M.J. Astle, eds. (1980) *CRC Handbook of Chemistry and Physics*. CRC Press, Boca Raton, FL.

Wright, S.H., T.W. Secomb & T.J. Bradley (1987) Apical membrane permeability of *Mytilus* gill: influence of ultrastructure, salinity and competitive inhibitors of amino acid fluxes. *J. Exp. Biol.* 129:205–230.

Yancey, P.H. & G.N. Somero (1979) Counteraction of urea destabilization of protein structure by methylamine osmoregulation compounds of elasmobranch fishes. *Biochem. J.* 183:317–323.

6 Hyporegulators

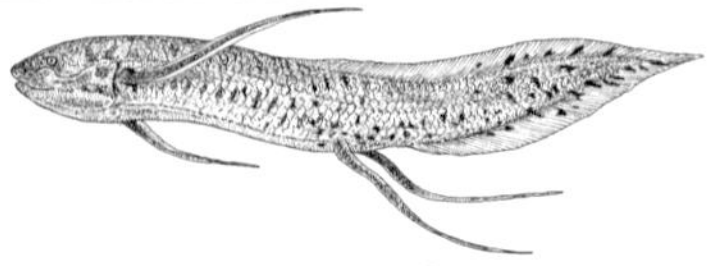

6.1 Introduction

As described in the previous chapter, the vast majority of marine animals, both in terms of the number of individuals and the number of species, are osmoconformers. This is a logical strategy for dealing with the marine environment as it equilibrates the activity of water inside and outside, thereby eliminating any substantial expenditures for water regulation. Yet a number of animal groups thrive in saline waters without being osmoconformers. These animals are all hyporegulators. They regulate the internal concentration of solutes at a level lower than the external medium. This chapter describes the physiological mechanisms employed by these animals.

6.2 General principles

Hyporegulators, by definition, maintain the osmotic concentration of their body fluids at a concentration below that of the external medium. Because the activity of water in the body is higher than that of the medium, the animals tend to lose water to the medium by osmosis. If this effect were not mitigated, the animals would lose water and thus body volume indefinitely. This problem could be alleviated if the animals had access to fresh water or to food that is lower in osmotic concentration than that of their body fluids. For most marine hyporegulators, these two solutions are not constantly available. Therefore, animals must drink the seawater to replace losses in their body volume. The animals replace the water that they have lost by drinking seawater, with the obvious drawback that it contains abundant salts, particularly sodium chloride. Hyporegulators, therefore, almost uniformly have some mechanism for producing and secreting a hyperosmotic fluid. The mechanisms by which these hyperosmotic secretions are formed, and the anatomical locations of their formation will be the major topics in this chapter.

6.3 Marine teleosts

Teleosts are the bony fishes. These terms are used to distinguish this clade of fishes from other vertebrates such as sharks and rays that have largely cartilaginous skeletal elements. The teleosts are an extremely successful group of animals. It is estimated that there are more than 20,000 species of marine teleosts.

Marine fish maintain an internal osmotic concentration of around 250–300 mOsm. They regulate this internal concentration very closely and as the external osmotic concentration of the ocean is about 1000 mOsm, marine fish qualify as strict hyporegulators. The first issue we should consider is the strong osmotic gradient that this mode of regulation imposes across the integument of the fish. The skin of marine teleost is relatively impermeable. The skin is covered with dead epidermal cells that contain abundant keratin fibers. The multiple layers of cell membranes and mats of protein fibers impede diffusive pathways and reduce osmotic flow across the skin. It is worth mentioning that damage to the skin of marine fish due to wounds, abrasions, or infections imposes a substantial osmotic load on the fish and is therefore a serious physiological issue.

The gills of fish consist of numerous thin gill filaments stiffened by cartilaginous rods running along their length (see Fig. 6.1 and Fig. 7.1 in the following chapter). When the fish respires, it draws water into the mouth, referred to as the buccal cavity. Following inspiration, the mouth is closed and positive pressure is exerted by the hyoid muscle to force water across the gill filaments and into the opercular cavity. From there the water flows out through the opercular openings, returning to the external medium.

Naturally, a critical role for this flow of water is the facilitation of the exchange of respiratory gases across the surface of the gills, the respiratory epithelium of the fish. The exchange of oxygen and the gaseous form of carbon dioxide across the gill epithelium is passive. Oxygen must diffuse across the gill epithelium, across the extracellular space, across the capillary wall, and into the blood. Carbon dioxide must passively diffuse along this same path in the opposite direction. Efficient transfer of these gases, therefore, necessitates that the gill epithelium be very thin. In addition, the flow of oxygen would be severely reduced if extensive connective tissue, keratinized structures, or scales were present on the gill surface. The cellular architecture of the gills is, therefore, organized along very different lines from the skin.

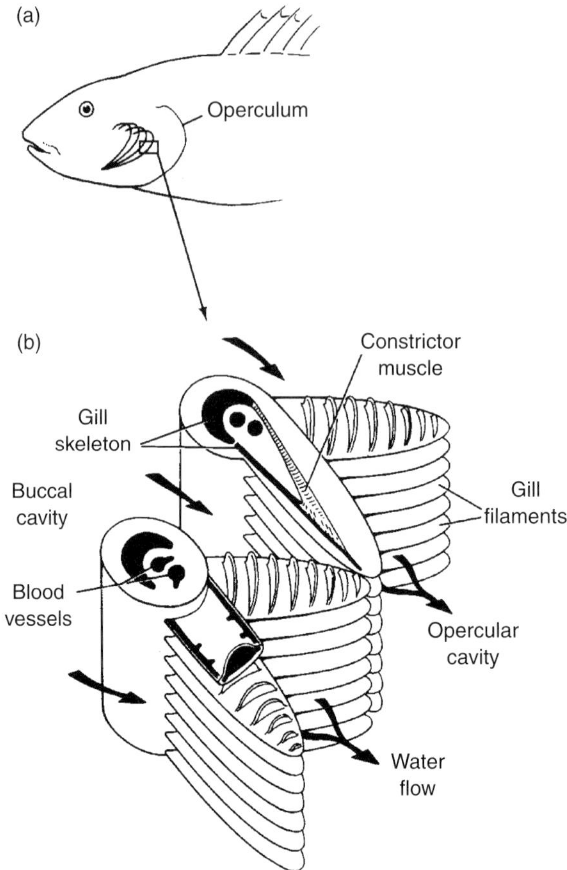

Fig. 6.1. An illustration of the pathway of water through the mouth and opercular chamber of fish. (Adapted from Schmidt-Nielsen, 1994.)

The thin epithelium covering the gills also is the major site of osmotic water loss from the body of marine fish. The necessity of having a repository surface that is highly permeable places an osmotic burden on the fish, given its hypo-osmotic condition relative to ocean water. The water lost across the gill is essentially pure water. As the gill membranes are relatively impermeable to ions, water can move out of the body across the gill epithelial cells and down its activity gradient, but ions move very slowly, if at all, down their activity gradient, which is directed inward.

The osmotically driven loss of water from marine fish would eventually be lethal if not countered. Loss of water without associated ions increases the osmotic concentration of the blood and cells, as well as reducing blood and extracellular fluid volume. The only means available to marine fish for replacing this water is drinking the external medium. The problem is that the seawater the fish must drink is contaminated with salts. In fact, as the seawater is more concentrated than the blood of the fish, the water they drink actually draws water out of the blood and into the gut. At this point, therefore, drinking seawater has led only to deleterious consequences.

As the ingested water enters the intestine of the fish, however, the ions are transported across the epithelium and into the interstitial fluid, eventually entering the blood. As these ions move across the gut, the activity gradient for water is modified and indeed reversed. By removing sodium, chloride, magnesium, calcium, and sulfate from the seawater in the gut, the fish moves the vast majority of the osmotically active solutes into that fluid. The activity of water in the gut fluid rises sufficiently that the water moves by osmosis across the intestinal epithelium and into the body fluids. At this point the fish has restored its body and blood volume, but it is still facing the problem of the excessive ionic load derived from the seawater.

The kidneys of most fish produce urine by filtering the blood through a filtration apparatus referred to as the glomerulus. The filtration barriers in the glomerulus filter out large, highly charged molecules such as proteins, but allow other fluid components to pass into the urine. This original filtrate is termed primary urine. Primary urine mimics the blood composition with regard to the small solutes present. It is rich in salts and small organic molecules.

Some marine fish lack glomeruli, but still produce primary urine. They do this by transporting ions, particularly magnesium, into the tubule lumen. Salts and water follow, producing urine that is isosmotic to the blood and rich in divalent ions.

To avoid eliminating the primary urine, which after all contains essential nutrients, the proximal tubules of the fish kidney resorb sodium, chloride, and organic compounds such as glucose and amino acids. The removal of these solutes increases the activity of water in the urine, and water diffuses osmotically across the proximal tubule cells and back into the blood. In addition, the proximal tubule cells actively transport divalent ions across the epithelium and into the urine. The result is urine which is isosmotic to the blood, low in sodium and chloride, and high in calcium and magnesium sulfate and bicarbonate. The urine is excreted in this form. Its excretion serves to remove the divalent ions that the fish has obtained by drinking seawater. The excretion of these solutes is very important in regulating ions that have profound physiological effects on nerve and muscle functions. However, the excretion of this urine does not solve the other osmotic problem associated with the drinking of seawater, namely, the ingestion of excessive amounts of sodium and chloride. Those problems are solved by the gill epithelium.

The epithelium covering the surface of the gills of fish is composed principally of three cell types (Fig. 6.2). The most numerous of these cell types are the pavement cells. As their name implies, pavement cells are thin and flat, by which I mean it is a relatively short distance from their apical surface to their basal surface. Similar to paving stones, the cells have interlocking margins and cover the surface of the gills. The short distance from the apical surface to the basal surface of the cells facilitates the rapid diffusion of oxygen into the gill and carbon dioxide out. Interspersed between the pavement cells are mucous cells that secrete mucous on to the surface of the gills. This mucous serves to protect the gills, particularly from potentially damaging particulate matter in the water stream, as well as from pathogens that stick to the mucous and are then sloughed off. The third cell type is the chloride cell, also sometimes called the mitochondria-rich cell. This cell type is the major site of ion transport across the gills. The chloride cells are much larger in their baso-apical axis. The cell contains numerous mitochondria. The basal membrane is greatly expanded in the form of folds and tubes that extend deep into the cell almost to the apical surface.

The chloride cells of marine fish are the site of active outward transport of sodium and chloride across the gill epithelium. Our current understanding of the molecular processes involved in this transport are outlined in Chapter 10. For our current discussion, it is sufficient to point out that the transport of sodium and chloride across the chloride cells occurs rapidly, the chloride cell has a sufficient low permeability to water, and that the

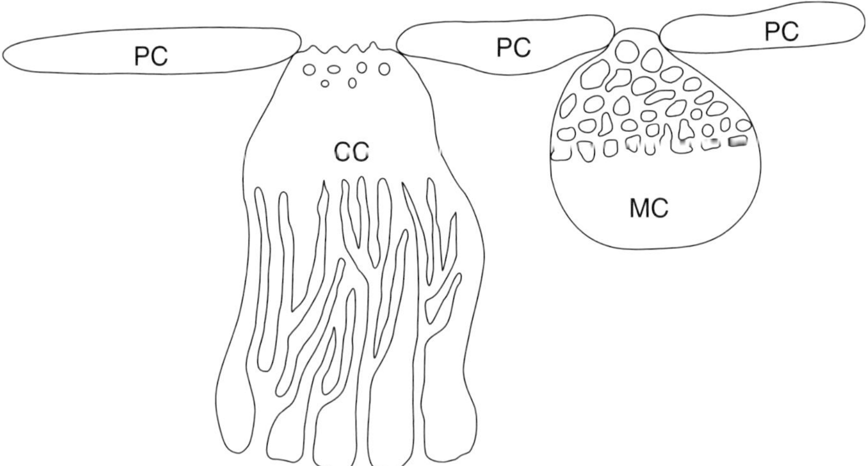

Fig. 6.2. There are three cell types in contact with the external fluid on the surface of fish gills. These are the thin pavement cells (PC); chloride cells (CC), which in marine fish are the site of salt excretion; and mucous cells (MC), which secrete protective mucous on to the gill surface.

transported fluid is hyperosmotic to both the blood and the seawater. The chloride cells of the gill, therefore, serve as the site of sodium and chloride excretion, and of the transport of a hyperosmotic fluid. These activities bring the fish back into osmotic homeostasis and allow the fish to maintain an osmotic concentration well below that of the seawater.

6.4 Marine reptiles

Several reptilian families contain species that inhabit the marine and/or estuarine waters. These include crocodiles, sea turtles, marine iguanas, and marine snakes. The vertebrates evolved from a fish-like ancestor. They have inherited and retained an osmotic concentration of their body fluids of around 300–350 mOsm. Marine reptiles are therefore hyporegulators.

Reptiles differ from both fish and amphibians in that they respire exclusively by means of lungs rather than gills or skin. As the lungs do not come into contact with the seawater, reptiles are able to have a respiratory epithelium that is thin and highly permeable to gases, but which does not lose water directly to the seawater. The skin of reptiles is thick and highly keratinized (see Chapter 9) for a further description of reptilian skin, characteristics that greatly reduce permeability to water. Nonetheless, no epithelium is entirely impermeable to water and marine reptiles still lose some water through their integument to the seawater. Thinner epithelia covering the mouth, nasal passages, and eyes are additional sites of osmotic water loss to the surrounding seawater.

Despite the relative impermeability of the integument of reptiles compared to fish and the isolation of the respiratory surface from contact with the seawater, marine reptiles still face a substantial osmotic load. If they feed on fish, they have food source with an osmotic concentration (300 mOsm) isosmotic to their own blood. It is likely, however, that they will ingest at least small quantities of seawater with the fish as they capture and eat them. Other food sources in the sea are osmoconformers as described in Chapter 5 and eating them is osmotically, although not ionically, equivalent to drinking seawater. (As the body fluids of osmoconformers contain substantial levels of organic solutes the ionic load is less than that of equal volumes of seawater.) This poses problems, for example, for crocodiles feeding on crabs, turtles feeding on jellyfish, or marine iguanas feeding on algae.

The reptilian kidney is incapable of producing urine more concentrated than the blood. In all of the vertebrates with exception of a few aglomerular fish, urine is produced by filtration in the glomerulus and modified in the kidney tubules. The primary urine produced in the glomerulus is isosmotic to the blood and thus markedly hypo-osmotic to seawater.

Marine reptiles maintain osmotic homeostasis by excreting a hyperosmotic, sodium chloride-rich fluid from their bodies into the surrounding waters. This capacity is carried out by epithelia in the head region. Crocodiles possess glands in the tongue that can secrete hyperosmotic fluid. In sea turtles, this activity is carried out by the lacrimal (tear) glands surrounding the eye. In marine iguanas, nasal glands are the site of active salt transport.

The situation with marine snakes is a bit more uncertain. Sea snakes also have skin of low permeability and breathe with lungs. Sea snakes feed on fish and as a result face lower salt loads than animals feeding on osmoconformers. The relative importance of hyperosmotic fluid secretion in sea snakes is still an unresolved issue for several families of marine snakes.

6.5 Marine birds

Similar to the reptiles from which they evolved, birds have skin that is highly impermeable to water. That coupled with a layer of feathers makes the rate of osmotic loss across the skin very low. Nonetheless, birds face hyperosmotic stress due to the loss of water via the respiratory tract and, in the case of marine birds, through the intake of hyperosmotic fluid when they drink seawater. If birds feed on fish, they are ingesting food in which the body fluids have an osmotic concentration of about 300 mOsm. If they are feeding on invertebrates such as crustaceans, mollusks, or annelids, then the food will have a concentration of about 1000 mOsm. Essentially all marine birds must deal with hyperosmotic stress.

Bird kidneys are capable of producing urine hyperosmotic to the blood. They do this using processes essentially identical to those of mammals, namely, the concentration of the urine in the collecting duct through the extraction of water into a hyperosmotic medulla of the kidney. A detailed description of this mechanism is provided in Chapter 8 under terrestrial mammals. For our purposes here, suffice it to say that the kidney of birds can produce urine with an osmotic concentration of about 700 mOsm, which is markedly hyperosmotic to the blood but still slightly hypo-osmotic to seawater. The urine, therefore, contributes to osmotic regulation but cannot fully achieve osmotic homeostasis on its own if the bird ingests seawater or prey isosmotic to seawater.

Marine birds use nasal salt glands to remove the remaining salts and maintain osmotic homeostasis. The glands actively transport sodium and chloride through a relatively impermeable epithelium. The resulting fluid is very concentrated, with an osmotic concentration that can be as high as 1300 mOsm (i.e., markedly hyperosmotic to seawater). It is worth noting

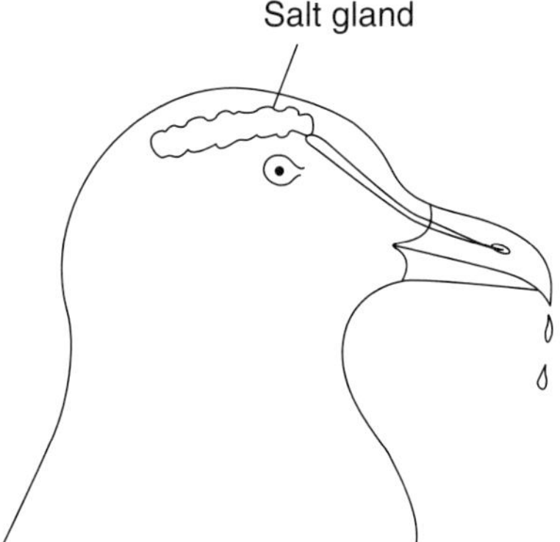

Fig. 6.3. Marine birds possess a cephalic gland that produces a hyperosmotic secretion. The fluid exits the bird through the nares.

that the salt glands in birds and reptiles are all in the form of glands in the head region.

The fluid secreted by the nasal glands exits the nares and flows out on to the beak (Fig. 6.3). If you watch seagulls at the shore you will often see them shake their heads to get rid of adhering drops of fluid on the bill. Those droplets are normally the product of nasal salt glands. Marine birds near land also avail themselves of inland sources of fresh water. By drinking at these sites, the birds presumably can reduce the energy output required for active ion transport in the nasal salt glands. Fresh water also seems to be a preferred medium for feather preening the birds engage in.

6.6 Marine mammals

All mammals are strict osmoregulators with body fluids that are maintained around 250–300 mOsm. Mammals arose on the land from reptilian ancestors. The mechanisms they employ for osmotic regulation were shaped by the terrestrial environment (see Chapter 8). Several lineages of mammals returned to the sea, including whales and dolphins, seals, otters, manatees, etc. They took with them their inherent mammalian characteristics and these were further shaped by the needs to meet their new environment.

Mammals have thick, keratinized skin covered with fur. The skin is relatively impermeable to water, a characteristic that aided mammals substantially in their evolved capacity to survive in the dry terrestrial environment. As a result, although marine mammals must survive in an aquatic environment markedly hyperosmotic to the blood, they lose relatively little water across the skin. In addition, mammals, as opposed to fish and most invertebrates, respire through lungs that come in contact only with the air; therefore, they do not face the problem of water loss to the marine environment through thin, highly permeable respiratory surfaces. On this basis, it might be argued that mammals face little osmotic stress in the marine environment. This is not entirely true, as you yourself would know if you were to fall overboard and drift at sea in tropical waters. In mammals, some water is lost in respiration. The significance of this process is described in more detail in Chapter 8. Marine mammals, however, breathe air that is immediately above the ocean surface and as such is usually very well hydrated. Respiration continues to be a small source of water loss in marine mammals, but it is a very small one. The principle osmotic stress on marine mammals comes from the ingestion of seawater and foods isosmotic to that medium. For animals feeding on marine invertebrates (e.g., otters feeding on clams, seals feeding on squid, or whales feeding on krill), the food is markedly hyperosmotic to their own blood. Even though a substantial portion of the osmotic activity in this food may be in the form of organic molecules, the mammals must nonetheless rid themselves of the excess ions in the food and deal with the drying effects of eating hyperosmotic food. Those mammals feeding on fish or other mammals (e.g., seals feeding on herring, killer whales feeding on seals) face a substantially smaller osmotic load but must nonetheless somehow replace water lost through evaporation from the lungs or osmosis with the surrounding waters.

Mammals solve all of these osmotic problems using their kidneys. The entire salt load and water loss experienced by marine mammals is countered by the production of concentrated urine. Being evolved from terrestrial mammals, marine mammals already possessed kidneys capable of hyperosmotic urine production (the mechanism for this is described in detail in Chapter 8). Upon entering the sea, mammals faced selection pressures for continued efficient and powerful kidney function. This capacity is impressive (e.g., it exceeds the urine concentrating capacity of humans) but not unprecedented, as a number of desert mammals have similar concentration capacities.

Ironically, some of the greatest osmotic stresses in marine mammals may arise when they exit the sea. Some seals spend a substantial portion of time on land, for example, to give birth to pups and to suckle them. These events frequently take place on islands with no shade. The seals are able to survive in this harsh terrestrial environment using only their kidneys as a means of producing a hyperosmotic excreta. The lactating females face the additional osmotic stress of producing dilute milk for the pups. The water they lose in this manner is a considerable osmotic stress. Seal milk tends to be extremely rich in fat, a mechanism to reduce water content and provide as many calories to the pup as possible with a minimum of water loss to the mother.

6.7 Saline-tolerant mosquitoes

Biologists have long been fascinated by the capacity of mosquito larvae to survive in saline environments that are much too toxic for any aquatic vertebrate. Although saline-tolerant, osmoregulating larvae are known to exist in several genera of mosquitoes, larvae in the genus *Ochlerotatus* have been the most carefully studied.

Larvae in the genus *Ochlerotatus* are strict osmoregulators. They regulate the osmotic concentration of their hemolymph at around 300 mOsm, regardless of the concentration of the external medium. As these larvae often are found in saline waters with osmotic concentrations well above that of seawater (>1000 mOsm), they clearly face large osmotic gradients driving water out of the larva. The cuticle of the larvae is permeable to water but quite impermeable to ions so the larvae constantly lose water to the external medium due to osmosis. This causes the hemolymph to increase in osmotic concentration and to be reduced in volume. The larvae replace the lost volume by drinking the external medium. This solves the problem of volume regulation, but the high salinity of the water exacerbates the problem of ion regulation.

As is the case with most insects, the Malpighian tubules are the site of urine production in mosquito larvae (Fig. 6.4). The tubules produce a fluid isosmotic to the hemolymph through the active transport of ions, principally potassium and chloride. This fluid flows into the gut at which point it moves posteriorly into the rectum. In freshwater forms, ions would be removed from this fluid to produce dilute urine (see Chapter 7). In saline-water species of *Ochlerotatus*, the rectum is divided into two parts, termed the anterior and posterior rectal segments. The anterior segment is the site of ion uptake and seems to be histologically and functionally similar to the rectum of freshwater species. The potassium and chloride ions that were used to form urine in the Malpighian tubules are resorbed in the anterior rectum.

For larvae residing in highly saline waters, the above mechanisms are inadequate for achieving osmotic homeostasis. The posterior rectum of the insects, however, restores osmotic balance by serving as a salt gland that produces hyperosmotic secretion, thereby removing the excess ions from the hemolymph. The posterior rectal segment actively transports sodium, magnesium, chloride, and potassium from the hemolymph to the gut lumen. The epithelium is relatively impermeable to water and the secretion produced can be as high as 3000 mOsm in osmotic concentration. The rectal fluid is always hyperosmotic to the medium in all waters in which

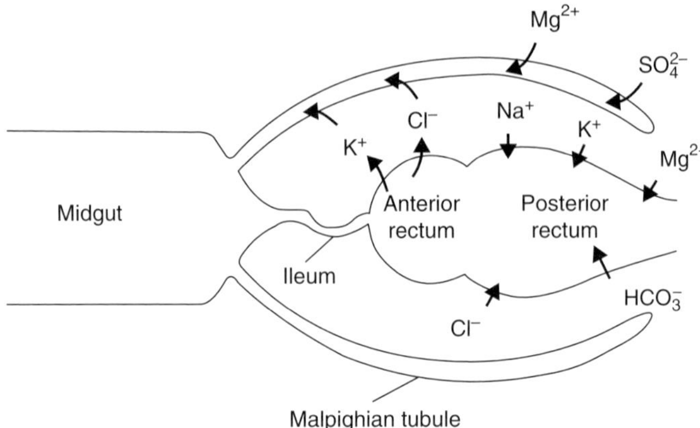

Fig. 6.4. A diagrammatic representation of the digestive and excretory organs of the larvae of a saline-water mosquito. Ions are actively transported into the Malpighian tubules, producing an isosmotic primary urine. This flows into the midgut and then posteriorly into the anterior rectal segment. Here, ions are resorbed. When the larvae are in concentrated media, the posterior rectum is the site of secretion of hyperosmotic fluid which is excreted through the anus. (Adapted from Bradley, 1985.)

the larvae can survive. The types of ions and rates of ion transport are influenced by the water in which the larvae had been reared. This allows the larvae to survive not only in seawater, but also in inland ponds rich in sodium bicarbonate or sodium and magnesium sulfate. No vertebrates are capable of surviving in these types of waters.

6.8 Brine flies

The larvae of Ephydrids (brine flies and shore flies) are astounding for their capacity to survive in highly saline waters. Brine flies from the Great Salt in Utah (USA) thrive by the billions in NaCl-rich water with a salinity up to 10 times that of the open ocean. At Mono Lake in California (USA) the larvae of *Ephydra hians* also thrive in water three times the concentration of the ocean, but these waters are rich in sulfate and bicarbonate, with a pH of 10!

All of the species of brine fly examined to date are osmoregulators. The hemolymph has a total osmotic concentration of about 300 mOsm. The osmotic concentration of the external medium can be well over 2000 mOsm. It follows that the larvae must lose water across the cuticle due to the large osmotic forces that exist. To replace the water lost through osmosis, the larvae drink the external medium. This poses the additional problem of forcing the larvae to rid themselves of the ions ingested.

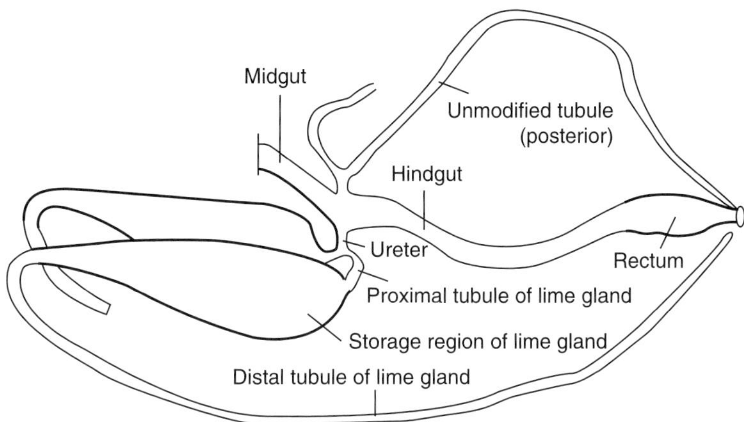

Fig. 6.5. A diagrammatic representation of the excretory organs of the brine fly *Ephydra hians*. (Adapted from Herbst and Bradley, 1989.)

Figure 6.5 illustrates the internal organs associated with the gut and with excretion in the larvae of *E. hians*. The fluids and food that are ingested pass into the midgut. The fluid in the midgut rapidly comes into osmotic equilibrium with the hemolymph. Water follows osmotically. The Malpighian tubules produce urine by transporting ions from the hemolymph into the tubule lumen. Water follows this transport osmotically as well. The fluid in the Malpighian tubules is, however, isosmotic with hemolymph and therefore does not contribute directly to osmoregulation.

Unlike mosquitoes, the rectum is not the site of hyperosmotic urine formation in *E. hians*. Instead, a section of hindgut anterior to the rectum termed colon is capable of producing a hyperosmotic fluid. The anterior half of the colon can actively transport sulfate, while the entire colon is capable of producing hyperosmotic urine rich in sodium and chloride. These transport processes permit the brine flies to be the most saline tolerant of all insect species.

6.9 Crustaceans (brine shrimp)

The champion hyporegulators, however, are brine shrimp. These crustaceans are found around the world in salt ponds and lakes. They can withstand extreme osmotic gradients and are frequently found in waters with precipitating salt crystals. In such waters, the brine shrimp often face no aquatic competitors or predators. Saline waters are generally rich in algae and bacteria on which the brine shrimp filter feed. Waters with abundant brine shrimp are often very attractive to birds that feed on the shrimp.

These birds, including phalaropes, grebes, gulls, and flamingos, often accumulate in spectacular numbers to feed on brine shrimp at inland salt lakes and ponds.

Brine shrimp are strict osmoregulators. They maintain their hemolyph at a concentration ranging from about 200 mOsm to 400 mOsm over a wide range of external salinities. The shrimp have a cuticle-covered integument that retards water loss, but the extreme gradients present still lead to substantial water and volume loss through osmosis. The shrimp drink the external medium to restore body volume. The accumulated salts are removed by powerful salt glands. In immature shrimp, termed nauplii, the gland is on the dorsal surface of the cephalothorax. In the adults, the second maxilla contains maxillary glands that are rich in Na^+/K^+ ATPase and are capable of transporting sodium chloride outward against steep gradients. The gut is also thought to be a site for the production of concentrated excreta.

Brine shrimp can survive periods of complete drying in the ponds in which they occur by means of cysts that the females produce and release into the water. These cysts have extremely low metabolic rates and can survive drying. Upon return to saline water, the cysts hatch, releasing the nauplii. The wading birds that are the major predators on brine shrimp play another important role in the life history of the shrimp. The cysts occasionally adhere to mud on the feet of these birds and in this manner are transported from one body of water to another. This means of transport carries the shrimp to new habitats and/or leads to gene flow between shrimp populations.

Suggested additional readings

Abatzopoulos, T.J., J.A. Beardmore, J.S. Clegg & P. Sorgeloos (2002) *Artemia Basic and Applied Biology*. Kluwer Academic Publishers, Dordrecht, The Netherlands.

Beyenbach, K.W. & P.L-F. Liu (1996) Mechanism of fluid secretion common to aglomerular and glomerular kidneys. *Kidney Intl*. 49:1543–1549.

Bradley, T.J. (1985) The excretory system: structure and physiology. In: *Comprehensive Insect Physiology, Biochemistry and Pharmacology*. G.A. Kerkut & L.I. Gilbert, eds. Pergamon Press, Oxford, NY.

Bradley, T.J. (1987) The physiology of osmoregulation in mosquitoes. *Ann. Rev. Ent*. 36:439–462.

Evans, D.H., ed. (2000) *The Physiology of Fishes*. CRC Press, Boca Raton, FL.

Herbst, D.B. & T.J. Bradley (1989) A Malpighian tubule lime gland in an insect inhabiting alkaline salt lakes. *J. Exp. Biol*. 145:63–78.

Hiroi, J., S.D. McCormick, R. Ohtani-Kaneko & T. Kaneko (2005) Functional classification of mitochondrion-rich cells in euryhaline Mozambique talapia (*Oreochromis mossambicus*) embryos, by means of triple immunofluorescence staining for Na^+/K^+-ATPase, $Na^+/K^+/2Cl^-$ cotransporter and CFTR anion channel. *J. Exp. Biol.* 208:2023–2036.

Ortiz, R.M., C.E. Wade, D.P. Costa & C.L. Ortiz (2002) Renal responses to plasma volume expansion and hyperosmolality in fasting seal pups. *Am. J. Physiol.* 282:R805–R817.

Schmidt-Nielsen, K. (1994) *Animal Physiology: Adaptation and Environment.* Cambridge University Press, Cambridge, MA.

7 Hyper-regulators: life in fresh water

7.1 Introduction

We will now consider a very important and widespread habitat: fresh water. Freshwater habitats exist nowhere in the ocean because if rain should fall on this massive accumulation of salt water, the dilution has no significant effect due to rapid diffusion and mixing. A possible exception to this broad statement is the effect of rain on tide pools and estuaries, but we discussed this already in Chapter 5.

Therefore, when we talk about freshwater habitats, we are discussing the aqueous habitats that occur on land. This includes streams, lakes, ponds, rivers, pools, swamps, and puddles, a wide array of hydrological forms. These habitats can differ widely in size, temperature, speed of flow, and opacity, but they are all surprisingly similar in their osmotic effects on animals.

The reason has to do with the general uniformity of osmotic conditions inside the animals themselves. All animal cells require specific quantities of potassium, chloride, phosphates and magnesium, sugars, amino acids, proteins, and nucleotides. In short, there is a laundry list of required substances in cells without which life simply cannot exist. For most organisms, this list of osmotically active compounds adds up to a total osmotic concentration of around 200–300 mOsm.

What then is the range of osmotic concentrations that are found in fresh water? At the lower end of the scale is rainwater (Table 7.1). Rain is, of course, distilled water formed by condensation in the atmosphere. There is a possibility of rain picking up a few salts from dust particles in the air, but on average rain has an osmotic concentration well below 1 mOsm.

Table 7.1. Salinity values

Water source	*Salinity (ppm)*	*Milliosmoles*
Rain (inland)	<3	<0.1
Rain (near coast)	~5	0.1
Lake in a well-watered landscape	11	0.3
Freshwater lake with a large drainage basin	46	1
St. Lawrence River (Canada)	50	1
Rhine River (Germany)	250	7
Murray River (Australia)	319	9
Colorado River (USA)	850	24
Fluid from the tissues of most freshwater animals	7,000–12,000	200–300
Seawater	35,000	1,000

When rain strikes the ground it picks up salts and organic compounds from the soil. Mountain streams have osmotic concentrations around 1 mOsm. As the water flows along streambeds or accumulates in lakes, the salinity can again rise by leaching from the soil and by atmospheric deposition of dust. As shown in Table 7.1, rivers can range in osmotic concentration from 1 to 25 mOsm. The Colorado River in North America and the Murray River in Australia are both so saline that they have limited value for irrigation. Table 7.1 shows the highest recorded osmolalities for these two rivers. But even these rivers, which are widely viewed as being highly saline, are well below the osmotic concentration of the intracellular fluids of any animal.

As a result, all freshwater organisms are faced with two physiological challenges. The first challenge is that water will diffuse down its activity gradient and enter the animal's body across the integument (skin, gills, body wall). This will occur in all freshwater organisms regardless of phylogenetic position. The second challenge is the paucity of necessary ions. Because freshwater is so low in ions, it provides a poor source of the sodium, calcium, potassium, and chloride required by all organisms. The physiological mechanisms employed by freshwater organisms to meet these two challenges is summarized in Table 7.2 and will provide the central theme to our discussions in this chapter.

7.2 Freshwater fish

We will begin our examination of the physiological responses to life in fresh water by examining the group of organisms that has been most intensively studied in this regard, freshwater fish.

Table 7.2. Mechanisms employed by freshwater organisms to meet the physiological challenges

Animal group	*Integumental features exposed to freshwater*	*Site of urine production*	*Extrarenal site(s) of ion uptake*
Freshwater fish	Gills, skin	Kidneys	Mitochondria-rich cells in the gill
Adult frog	Skin	Kidneys	Skin
Salamanders	Gills, skin	Kidneys	Gills
Freshwater turtles	Skin	Kidneys	Rectum
Freshwater snakes	Skin	Kidneys	None known
Freshwater otters	Skin	Kidneys	None known
Beavers	Skin	Kidneys	None known
Mosquito larvae	Cuticle-lined external integument	Malpighian tubules	Anal papillae
Mayfly larvae	Cuticle-lined external integument	Malpighian tubules	Anal gills
Earthworms	Skin	Nephridia	Skin
Freshwater crabs	Cuticle-lined external integument	Green gland	Gills
Amoebae	Plasma membrane	Contractile vacuole	Plasma membrane
Freshwater clam	Gills, mantle, surfaces of numerous organs	Nephridia	Gills, mantle

Most fish, whether residing in fresh water or salt, have tissue osmotic concentrations of about 230–300 mOsm. The extracellular fluids are rich in sodium and chloride, while the intracellular fluids are rich in potassium, magnesium, and organic compounds. As indicated in Chapter 5, the intracellular and extracellular fluids are isosmotic throughout the body.

As a result, freshwater fish are faced with a substantial inward-directed gradient for the activity of water, and a large outward-directed gradient for ions, particularly sodium and chloride. Given these gradients, a major physiological issue is the permeability of the external surface (integument) of the fish to ions and water. As indicated in Chapter 2 (Table 2.1), the integument of the fish has a finite permeability to water and ions; in other words, the animals are unable to make themselves completely impermeable to either ions or water. However, the surface of the animal is not uniform with regard to these parameters. The skin of the fish, that is, the surface of the body covering the head, flanks, and fins, is relatively impermeable to water. The skin of fish is a complex, squamous epithelium. This means that the epithelium is composed of multiple layers of flattened cells, resulting is a thick layer of cells. The diffusive pathway across this epithelium traverses many layers of cellular membranes. The resistance of these membranes is additive, leading to a very low overall permeability of the entire epithelium. This serves the purpose of the fish very well, because the skin is meant to serve as a barrier and not as a conductive epithelium.

In contrast to the skin, the epithelium covering the gills is much more permeable (Table 2.1). Let us examine the cellular architecture of the gill epithelium of fish to understand why this might be the case.

7.2.1 The structure of the gills of freshwater fish

The gills of fish consist of numerous thin gill filaments stiffened by cartilaginous rods running along their length. When the fish respires, it draws water into the mouth, referred to as the buccal cavity. Following inspiration, the mouth is closed and positive pressure is exerted by the hyoid muscle to force water across the gill filaments and into the opercular cavity. From there the water flows out through the opercular openings, returning to the external medium (Fig. 7.1). The general organization of the gills has been shown in Chapter 6 (see Fig. 6.2).

Naturally, a critical role for the gills is the exchange of respiratory gases between the blood and the external aqueous medium. The exchange of oxygen and the gaseous form of carbon dioxide across the gill epithelium is passive. Oxygen must diffuse across the gill epithelium, across the extracellular space, across the capillary wall, and into the blood. Carbon dioxide must passively diffuse along this same path in the opposite direction. Efficient transfer of these gases, therefore, necessitates that the gill epithelium be very thin, and that fluid flows rapidly adjacent to a large epithelial surface area (Fig. 7.2). The flow of oxygen would be severely reduced if extensive connective tissue, keratinized structures, or scales were present on the gill surface. The cellular architecture of the gills is, therefore, organized along very different lines from the skin.

The epithelium covering the surface of the gills of fresh water fish contains pavement, chloride and mucous cells (Fig. 7.2). The most numerous of these cell types are the pavement cells. As their name implies, pavement cells are similar to paving stones; the cells have interlocking margins and cover the surface of the gills. Mucous cells that secrete mucous on to the surface of the gills. The third cell type is the chloride cell, also sometimes called the mitochondria-rich cell. This cell type is the major site of ion transport across the gills.

It can be seen that the cellular architecture required for efficient transfer of respiratory gases leads to a circumstance in which a highly permeable epithelium is brought into constant contact with rapidly flowing water. As a result, freshwater fish experience a rapid inward flux of water across the

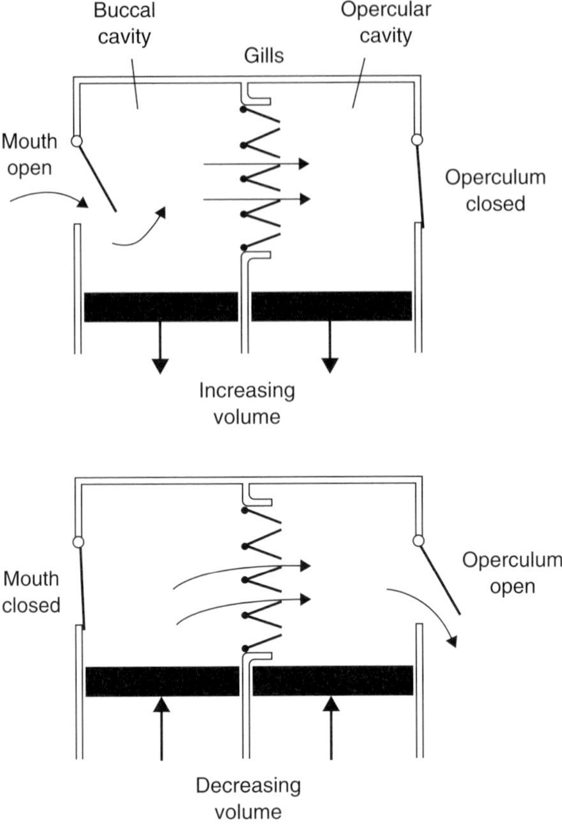

Fig. 7.1. A functional model of water movements through the mouth, oral cavity, opercular cavity, and operculum of a fish. (Redrawn from Randall *et al.*, 1997.)

gill epithelium. This influx dilutes the internal fluids of the fish, threatening its survival. The physiological response of the fish is the production of dilute urine. As indicated in Chapter 6, marine fish sometimes have aglomerular kidneys. This is not the situation in freshwater fish. Instead, freshwater fish always possess glomeruli and these are the site of primary urine formation (Fig. 7.3). The kidneys of freshwater fish produce urine by filtration. This means that the blood pressure in the capillaries of the glomerulus is used to drive fluid across the glomerular walls into Bowman's capsule and ultimately into the proximal tubules. As the primary urine is produced by filtration and not by active transport, the primary urine is isosmotic with the blood and similar to the blood plasma with regard to the principal small dissolved solutes (salts, small organic sugars, organic acids, etc.) present.

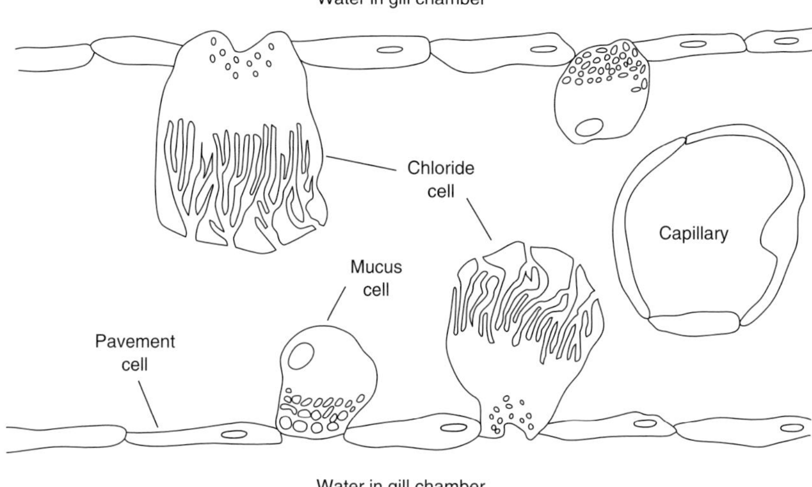

Fig. 7.2. An illustration of the three cell types in the fish gill and their positions relative to the water flowing across the gill.

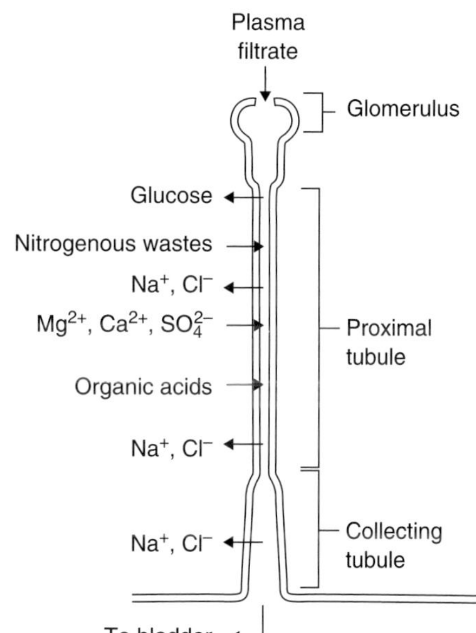

Fig. 7.3. The organization of a kidney tubule in a freshwater fish and the site of transport of compounds across the tubular epithelium.

The primary urine formed in the glomerulus is captured in Bowman's capsule and flows by hydrostatic pressure into the proximal tubules (Fig. 7.3). The epithelial cells lining the proximal tubules are capable of rapid ion transport, returning sodium, potassium, and chloride to the blood. In addition, the proximal tubules possess transport mechanisms for returning organic compounds to the blood, including glucose, amino acids, and vitamins. The result is the removal of a large percentage of the osmotically active compounds in the urine and the production of dilute urine. To retain this water in the urine, the kidney tubules must be relatively impermeable to water. This is indeed the case. The inner (luminal) surface of the kidney tubule epithelium is ciliated to promote a rapid flow of fluid along the tubule. The urine is rapidly transported through the tubules and into the bladder, which also has a low osmotic permeability. The fish are therefore able to excrete the water that enters across the skin, and even more rapidly across the gills, through the production of dilute urine.

Even though the urine produced is very dilute, it never can be entirely free of ions. The lowest measured values for the osmotic concentration of urine from freshwater fish were 20 mOsm. Fresh water is generally much lower than the urine with regard to total osmotic concentration, as well as sodium and chloride concentrations. It follows that the process of forming dilute urine, while it might permit the fish to maintain volume homeostasis, cannot maintain osmotic and in particular ionic homeostasis. We must now, therefore, consider the role of ion transport in freshwater fish.

7.2.2 Ionic homeostasis

In freshwater fish, the permeability of the skin with regard to ions is extremely low and can in fact be largely ignored. The gills, however, have a measurable, passive permeability to ions due to the very thin and delicate epithelium that covers the respiratory surfaces.

As pointed out above, in the absence of active processes the fish would eventually die of ion depletion. The gills of fish, however, are also the site of powerful active transport processes that serve in the uptake of sodium and chloride from the external medium. The ionic gradients for both these ions are steep, with the internal concentration being two to three orders of magnitude greater than that of the external concentrations. The uptake of these ions must therefore be active and is coupled to the active use of metabolic energy to do thermodynamic work. For our current discussion, suffice to say that sodium and chloride are actively absorbed up from the medium to replace the ions the fish loses in the urine. A more detailed description of our current understanding of ion transport in fish gills is provided in Chapter 10.

The process of ion uptake in the gills of freshwater fish serves another very useful function, namely, the removal of metabolic wastes from the body. When sodium ions (which are positively charged in solution) are

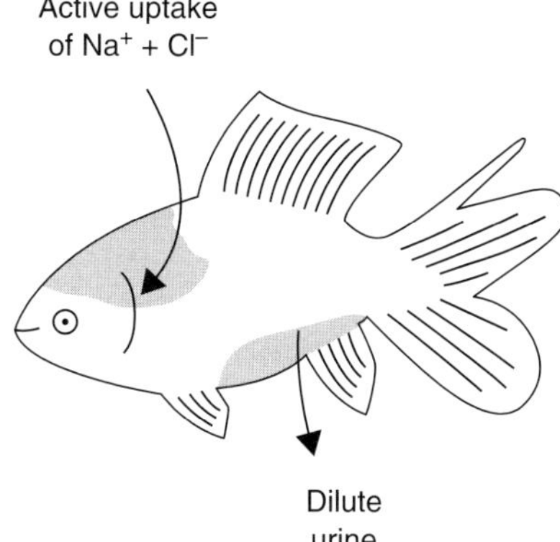

Fig. 7.4. Sites of ion uptake and water removal in a freshwater fish.

taken up across the gills, this is coupled with the excretion of hydronium and/or ammonium ions (which are also positively charged). This process serves to greatly reduce the energy needed for the active transport as the outcome of the exchange is electrically neutral. (See Chapter 9 for discussion of the influence of ionic charge on thermodynamic work and ion transport.) Similarly, the uptake of Cl^- ions occurs simultaneously with the excretion of negatively charged bicarbonate ions derived from CO_2. As a result, the uptake of sodium and chloride by the gills is facilitated by the excretion of two waste products: ammonia derived from the deamination of proteins, and bicarbonate derived from carbon dioxide. Other vital ions such as calcium, magnesium, potassium, and others are obtained largely from the food.

In summary, in order to deal with the osmotic entry of water, freshwater fish produce a urine equal in volume to the amount of water entering (Fig. 7.4). Ion balance is achieved by making the urine as dilute as possible. In very dilute waters this process is inadequate to maintain osmotic balance and the fish must also take up sodium and chloride across the gills. This process is energetically costly but is necessary for survival in freshwater.

7.3 Amphibians

7.3.1 Adult frogs

Adult frogs are essentially aquatic animals, despite their occasional forays into the terrestrial realm. Unlike the skin of fish, the skin of frogs serves as a respiratory surface. Respiratory organs must by their very nature be relatively permeable and highly vascularized. As a result of these characteristics, the skin of frogs is a site of rapid water loss to subsaturated air. The kidneys of amphibians, similar to those of the fish-like ancestors from which they evolved, are incapable of producing urine more concentrated than the blood of animals. Therefore, although frogs may be capable of spending considerable time on land, they must regularly return to an aquatic habitat to restore the water lost from the body. In this regard, frogs and toads possess an interesting physiological adaptation for replacing water lost by evaporation to the terrestrial environment. The skin on the belly and hind legs of these amphibians (referred to as the pelvic patch) has a particularly high osmotic permeability. Therefore, in order to achieve water uptake, the amphibians need not be fully immersed in water, they need only to sit in a puddle allowing the eyes and nostrils to remain in the terrestrial realm. In fact, both frogs and toads have been shown to take up water osmotically across the pelvic patch even when they sit on a wet substrate such as mud. This adaptation is vital for allowing these essentially freshwater animals to temporarily invade the terrestrial environment for the purposes of migration and foraging.

All amphibians are tightly linked to the aquatic environment both by their reproductive requirements, and by their osmotic limitations. When immersed in water, the high permeability of the frog skin leads to osmotic water influx. The kidneys respond by producing a copious, dilute urine. As in fish, this urine can be made dilute by the active retrieval of ions from the primary urine, but the urine is never as dilute as the freshwater in which frogs usually reside.

We are by now familiar with the solution to this problem, frogs must have a mechanism for replacing these lost ions. Unlike fish and tadpoles (see Box 7.1), however, frogs lack gills. When immersed, frogs respire by exchanging gases across the skin. This provides the animals with a large surface area for exchange. When the frog has access to air, the lungs are the principal organs for gas exchange, although the skin continues to provide additional respiratory support.

As adult frogs lack gills, this pathway for ion uptake is not available to them. The capacity for rapid ion uptake in frogs has been transferred to the skin. The capacity of frog skin to transport ions has been known for over 120 years. Frog skin is a relatively tough epithelium, compared to gills or guts. In addition, it is capable of supplying its own respiratory support as it is itself a respiratory organ. For these reasons, frog skin has been the focus of numerous studies examining ion transport in isolated tissue (referred to as *in vitro* studies, Latin for isolated tissues studied *in glass*). These investigations have demonstrated that frog skin contains powerful ion transport processes. The skin is capable of actively transporting sodium and chloride from the medium to the blood at a rate

Box 7.1. Tadpoles

In considering the processes used by amphibians in fresh water, we should also consider the developmental stage of anurans (frogs and toads) that is wholly aquatic, namely, the tadpoles. Tadpoles, the immature larval forms of frogs and toads, face osmotic circumstances similar to those experienced by freshwater fish. The internal fluids of the tadpoles range in osmotic concentration from about 200 mOsm to 300 mOsm. As we had noted previously, the freshwater environment is substantially more dilute. Unlike fish, the skin of amphibians generally has a relatively high osmotic permeability (Table 2.1). As in fish, the surface of the gills of tadpoles consists of very thin epithelia appropriate to the role of facilitating gas exchange. As a result, the gills are also very permeable to water and, to a lesser extent, ions. Unlike fish, the skin of amphibians is much less keratinized and lacks scales. As a result, the skin of amphibian is a more significant route for osmotic water entry than the skin of fish.

Tadpoles are faced, therefore, with a substantial rate of water uptake by diffusion across both the gills and the skin. They excrete this water via the kidney through the production of dilute urine. Amphibians posses kidneys similar to those of fish with well-developed glomeruli. Urine is formed by filtration of the blood plasma and the kidney tubules serve to remove essential ions from the primary urine. The bladder also serves as a site of ion uptake from the urine. However, as in fish, the urine produced by the tadpoles cannot be made as dilute as the waters in which the animals reside. The gills of tadpoles possess cells responsible for the uptake of ions (principally sodium and chloride) from the aquatic medium. These processes serve to restore ionic and osmotic balance.

that replaces ions lost in the urine. Frogs, therefore, solve the problem of ion replacement in the same manner as most other freshwater animals, namely, by using an extrarenal organ for ion uptake. In the case of frogs, this organ is the skin. Naturally, ions obtained in the food (principally terrestrial insects and mollusks) are also very beneficial in restoring ionic balance in this freshwater organism.

7.3.2 Salamanders

We should deal briefly with salamanders, another group of amphibians that has been extensively studied with regard to osmoregulation. Salamanders are entirely dependent on external water resources and are incapable of producing urine more concentrated than the blood. The kidneys of salamanders have been the subject of extensive study because they possess large glomeruli that are easily visible when they lie near the surface of the exposed kidney following dissection. For these reasons, salamanders were the subjects of numerous studies in the 1950s and 1960s era when kidney tubule micropuncture provided tremendous insights into kidney function. It was determined that the kidneys of salamanders have large glomeruli that promote the rapid formation of primary urine. The kidney tubules can remove ions from this fluid, but are incapable of producing urine that is hyperosmotic to the blood.

Even those salamanders that are mostly terrestrial (lunged and lungless salamanders) must reside in habitats where the activity gradient for water favors substantial osmotic influx of water. Examples of such habitats include wet forest floors, cavities under logs, sites saturated with water vapor, or habitats that permit periodic access to water (such as stream banks) for osmotic water loading. When in water, salamanders absorb water across the skin and/or across the gills if these are present. The skin and gill epithelia are both capable of active sodium and chloride uptake. The salamanders therefore employ physiological mechanisms identical to those of other amphibians with the exception that the sites of ion uptake vary depending on the presence or absence of gills in the various developmental stages and/or species.

7.4 Reptiles

Reptiles are distinguished from the amphibians in part by their highly differentiated integument. The skin of reptiles is highly keratinized, thick, and often covered with scales. This tough, impenetrable exterior has allowed reptiles to invade a number of habitats unavailable to amphibians, including terrestrial and marine habitats. Reptiles also are found in freshwater. In such environments, the reduced permeability of their integument greatly reduces the osmotic uptake of water that we have discussed at length in the fish and amphibians. Similarly, the evolution of lungs has allowed them to expose their highly permeable respiratory epithelia to air alone, and not to the external freshwater environment. As a result, reptiles inhabiting freshwater have much reduced osmotic loads.

Despite their relative integumental impermeability, reptiles do experience some water loading while residing in freshwater. Some water does diffuse across the skin, albeit very little. In addition, residing in freshwater and catching prey in this environment inevitably leads to the ingestion of freshwater, either intentionally or as a bycatch of prey capture. As a result, freshwater reptiles such as turtles, water snakes, and alligators obtain ample water for urine production. The urine produced under these conditions is more dilute than the blood of the animal, but once again more concentrated than the external medium. Habitation in freshwater environments brings with it the difficulty of obtaining sufficient ions for physiological functions. The principal source of these ions in aquatic reptiles is their food. By ingesting ions in the food and minimizing water uptake, reptiles are able to achieve osmotic balance in the freshwater environments.

Despite the general adequacy of the reptilian strategy of relying on ingested ions for ionic balance, at least one reptilian group does possess an extrarenal site of ion uptake from freshwater environments. Freshwater turtles can actively transport ions across the gut wall, including the rectal epithelium. Several species of turtle can also open the anus and perfuse the rectum with water from the external environment. This provides the rectal epithelium with oxygenated water and has been shown to be an important source of oxygen for inactive turtles when submerged under water, for example, during long dives where the animal rests on the bottom. That same water flowing across the rectal epithelium provides ions that can be transported across the rectal epithelium and into the blood.

7.5 Birds and mammals

Despite their many anatomical and physiological differences, birds and mammals can be considered together with regard to their osmotic strategies in freshwater. Both taxa have skin that is highly impermeable to water. The integumental epithelia are highly stratified and covered with multiple layers of dead flattened cells that greatly reduce the diffusion of water across the integument. In addition, many of the animals also secrete lipids on to the skin to further reduce the diffusion of water and heat loss. All birds and mammals are capable of forming dilute urine. In this manner, they rid the body of water obtained by ingestion and conserve ions to the greatest extent possible. Acquisition of the ions necessary to make good the loss in the urine occurs exclusively through ingestion. No extrarenal sites capable of ion uptake directly from the external environment occur in either birds or mammals.

7.6 Insects

7.6.1 Mosquito larvae

Unquestionably, among the aquatic insects, the most intensively studied with regard to osmotic regulation are mosquito larvae. This is true not only because of their ubiquity around the globe, but also because of the critical role that adult mosquitoes play in spreading deadly human diseases.

Mosquito larvae are found in almost every kind of freshwater habitat in which predators upon them are lacking or occur in low numbers. These include the tropical, temperate, and arctic regions, and habitats of varying sizes, ranging from collections of water in leaf axils to marshes stretching for kilometers across a flooded plain. As you can imagine, the osmotic characteristics of these habitats differ substantially, with rain-filled bromeliads and leaf axils high in the tropical canopy being probably at the low end of the osmotic scale, and muddy puddles drying in the sun being much higher in osmotic and ionic concentrations. Mosquitoes can survive in all these habitats, although the species present will vary depending on the geographical location and the physical parameters of the aquatic habitat.

The entire larval phase of the mosquito life cycle is spent in water. Mosquitoes are holometabolous, meaning that the larvae (Fig. 7.5) are morphologically distinct from the adults. Similar to all freshwater organisms, mosquito larvae must maintain an osmotic concentration in their extracellular fluid, which is much higher in dissolved components than the external medium. In the case of insects, the circulating extracellular fluid is not confined to vessels in the circulatory system. Instead, the fluid flows gently about in an open circulatory system. This fluid is referred to as hemolymph. The osmotic concentration of the hemolymph in most freshwater species ranges from about 200 mOsm to 270 mOsm. As a result, the hemolymph is much more concentrated than the external freshwater medium.

At this point, we need to consider the rate of uptake of water by osmosis in these animals to determine the rate at which water must again be eliminated. The integument of insects is covered by cuticle, a thick layer of nonliving material secreted by the underlying epidermal cells. Cuticle is a composite material composed of proteins and a fibrous carbohydrate termed chitin. By linking the proteins and chitin fibers into a tough, elastic layer, insects can protect their external surfaces from abrasion and pathogens, while simultaneously reducing the osmotic permeability of the integument.

As we will read subsequently in our discussion of the osmotic relations of terrestrial animals (Chapter 8), insects can make their integuments quite impermeable to water through the secretion of an external, superficial layer of wax upon the cuticle. This elegant physiological solution is in general not available to aquatic insects. A layer of wax would make the insects'

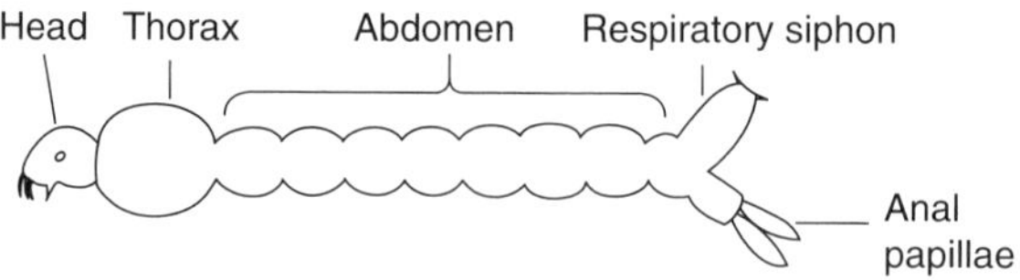

Fig. 7.5. Aspects of the external morphology of a mosquito larva.

surface nonwettable, and the insect would as a result stick to the air-water interface at the surface of the water. The hydrophobic forces at this surface are extremely strong, particularly for small insects that have a huge surface to volume ratio. Therefore, aquatic insects generally have only very small regions of nonwettable waxy coating. In mosquito larvae, for example, the tip of the respiratory siphon is nonwettable to promote a strong, stable connection between the tracheal system and the atmosphere above the water surface. As a result, mosquito larvae have no mechanism for making the cuticle completely impermeable to water.

Mosquito larvae feed on microorganisms in the water, ingesting suspended bacteria, protozoa, and algae by filtering the water using cuticular combs around the mouthparts. Alternatively, they can use their mandibles to graze on attached algae. Consumption of these types of food leads to the simultaneous ingestion of substantial volumes of water. Larvae ingest up to 12% of their body volume per day. As a result, mosquito larvae obtain substantial quantities of water osmotically across the cuticle by drinking and feeding. The resulting uptake of water is, on a per volume basis, much higher than that we observed in fishes, and necessitates that the insects must have a very robust mechanism for excreting water and obtaining the ions they require from a very dilute medium.

Ingested water passes along the gut and enters the hemolymph through the walls of the midgut when osmotically active solutes (ions and nutrients) are transported across the gut wall (Fig. 7.6). Water entering across the cuticle also ends up in the hemolymph. Both processes dilute the hemolymph. Urine formation in the insects occurs by mechanisms that are entirely distinct from those in the higher vertebrates, both with regard to mechanism and cell types. Malpighian tubules are the site of urine formation in most insects. These tubular epithelia are composed of a single layer

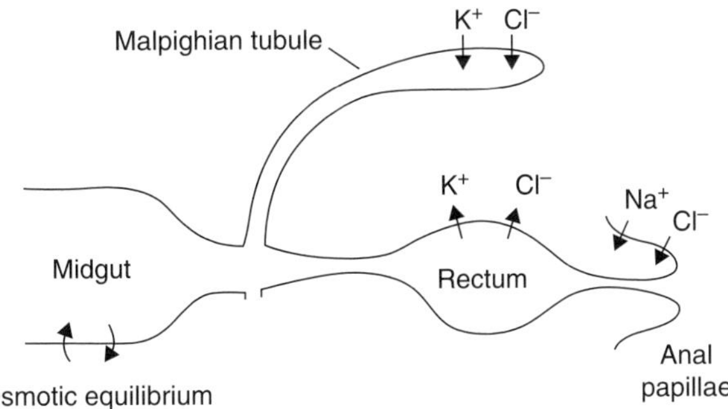

Fig. 7.6. Pathways of active ion movement in the osmoregulatory organs of mosquito larvae.

of cells. The cells transport ions from the hemolymph into the lumen of the tubules. The Malpighian tubules have a very high osmotic permeability, so the transport of osmotically active solutes leads to a rapid flow of water across the epithelial membranes. Whereas the urine in freshwater fish, reptiles, amphibians, birds, and mammals is produced by filtration using the hydrostatic pressure of the blood, the primary urine in insects is produced solely by osmotic flow driven by active ion transport. Owing to the high osmotic permeability of the tubules, the urine produced is essentially isosmotic to the hemolymph.

The Malpighian tubules serve to remove water from the hemolymph, thereby regulating hemolymph, and thus, body volume. The tubules serve no purpose with regard to osmotic regulation, however, as the primary urine produced is isosmotic to the hemolymph. The transformation of this primary urine into a dilute urine occurs in the hindgut. The copious fluid produced by the Malpighian tubules flows into the posterior portions of the midgut and then posteriorly into the hindgut. In the rectum, ions in the urine are rapidly retrieved and returned to the hemolymph by means of active transport across the epithelial cells in the rectal pad. In contrast to the Malpighian tubules, the epithelium of the rectum has a very low osmotic permeability. The removal of ions from the urine occurs with little water following. The resulting dilute urine is excreted via the anus.

As we observed with the vertebrates, mosquito larvae are capable of producing very dilute urine, but they cannot produce a fluid as osmotically dilute as the external medium. We can conclude that the insects must possess an extrarenal site for ion uptake. In fish and amphibians, this extrarenal site was also a respiratory organ, providing the advantage that the site of ion uptake was thin and permeable epithelium constantly perfused with blood on the basal surface and water on the apical. In mosquito larvae, respiration is achieved by means of air-filled tracheae that are open to the atmosphere when the larva is at the water surface. Such a respiratory system offers no opportunity for ion uptake from the medium.

In mosquito larvae, therefore, the extrarenal site of ion uptake bears no relation to the respiratory organs. Instead, it consists of sausage-like extensions on the posterior end of the animal termed the anal papillae. The cells of the anal papillae (which are a syncytium for aficionados of cellular architecture) are capable of active transport of sodium and chloride from the external medium into the hemolymph. Similar to the gills of freshwater fish, the cells of the anal papillae are capable of transporting ions against a large concentration gradient. The precise mechanisms of transport in anal papillae are still under investigation but it is clear that active chloride transport occurs in these organs. Mosquito larvae can maintain ionic balance in very dilute media even in the absence of food. Ablation of the anal papillae leads to a loss of ionic homeostasis in very dilute media, but not in more concentrated media in which the larvae can maintain homeostasis with renal mechanisms alone.

It is valuable to compare and contrast the osmoregulatory mechanisms of mosquito larvae with those of freshwater fish. Clearly, the ubiquity of mosquitoes argues for their capacity to inhabit the most dilute and ephemeral habitats available to freshwater organisms. They achieve this by forming urine and retrieving ions from this fluid. Nonetheless, urine can never be made as free of ions as the most dilute media and similar to fish, the insects require additional sites for ion uptake. The overall osmotic strategy in freshwater fish and freshwater insects is therefore identical, although the organs and cell types employed in these strategies are entirely distinct and arose through independent lines of evolution from primitive marine ancestors.

7.6.2 Mayflies and Dragonflies

The diversity of insects, which invaded the freshwater aquatic habitat, is impressive. Over 20 orders of insects have life stages that are strictly aquatic. Insects play a very important role in freshwater aquatic ecosystems as grazers, predators, and as food for secondary consumers, particularly fish. Let us consider two additional freshwater insects just to get a sense of the diversity of body plans and physiological processes employed by insects in surviving in hypo-osmotic environments.

Mayflies possess larvae that are restricted to freshwater habitats. The larvae feed on algae and detritus and, therefore, obtain little sodium and chloride in the diet. The larvae have a large surface area because of their elongate shape and the presence of three pairs of legs, cuticular hairs, and numerous gills arrayed along both sides of the abdomen (Fig. 7.7). The cuticle of mayflies is permeable to water and this leads to a constant influx of osmotically driven water. The osmotic permeability of the cuticle of mayflies is similar to that of mosquito larvae. The influx of osmotically driven water into the mayfly necessitates the production of dilute urine. This occurs in mayfly larvae in the Malpighian tubules and ion resorption takes place in the rectum.

Unlike mosquito larvae, however, mayfly larvae do not exchange gases directly with the atmosphere via a siphon, but respire instead using gills. As in fish and amphibians, the respiratory functions of gills dictate that a large and very thin surface area must be brought into intimate contact with fast-flowing external water in order to obtain oxygen and lose carbon dioxide at a rate sufficient to support oxidative metabolism. In mayfly larvae, the gills consist of leaf-like appendages (Fig. 7.8) attached to the lateral surfaces of the abdomen. The gills are relatively large and flat, providing a

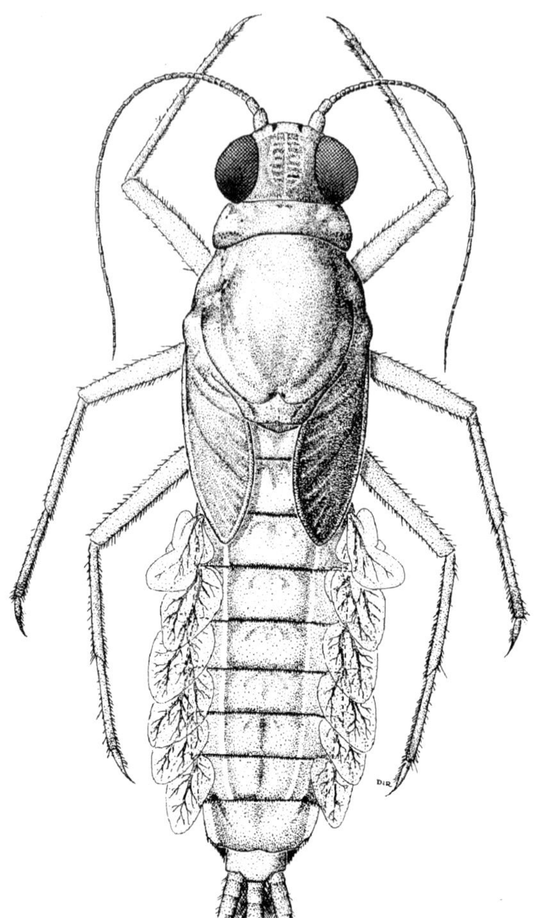

Fig. 7.7. A mayfly larva. Note the leaf-like gills located on the lateral edges of the abdomen. (Adapted from Edmunds et al., 1976.)

large surface area for gas exchange. Unlike the situation in the vertebrates, the gas exchange does not occur between the water and the blood containing hemoglobin. Instead, in the mayfly larvae the gas exchange occurs between the external water and air-filled tracheae that are highly ramified in the internal spaces in the gills (Fig. 7.8). The tracheae permit rapid diffusion of gases down their concentration gradients to and from the respiring tissues.

The gills of mayfly larvae possess specialized cells on their surface that are capable of transporting sodium and chloride ions against the concentration gradient that exists between the freshwater environment and the insect's hemolymph. The cells which carry on this transport are termed chloride cells as they were originally thought to be unusually rich in

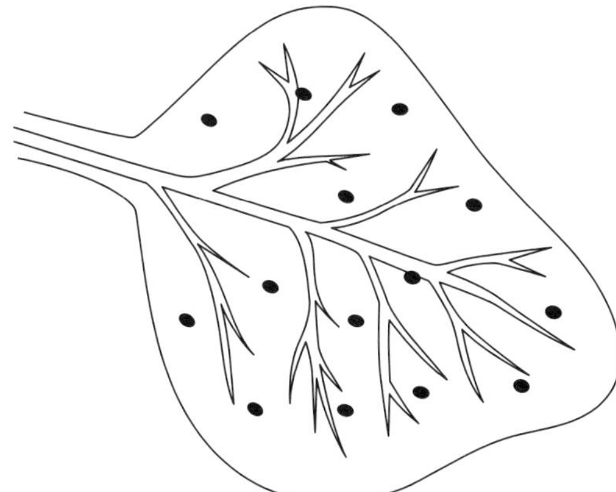

Fig. 7.8. A magnified image of a single mayfly gill. The dark circles are chloride cells, the sites of active ion uptake.

chloride due to their capacity for staining with silver stain. In fact, this silver staining did not elucidate unusual chloride concentration but rather sites of unusually high permeability in the cuticle, as befits a site where the underlying epithelial cells must have easy access to external ions.

It is interesting to note that the uptake of sodium and chloride in mayfly larvae occurs in the gills, just as is the case in fish. The physiological significance of ion uptake on the respiratory surfaces lies in the fact that the ions can be taken up from water that is flowing rapidly across the animal's integument. Although the flow of water is driven principally to prevent stagnation of the external medium and to facilitate gas exchange, the flow of water also supplies new ion-rich water for the purposes of ionoregulation.

7.6.2.1 Dragonfly larvae

Everyone is familiar with those beautiful and accomplished aeronautic specialists, the dragonflies. The adults are showy and their tendency to leave their aquatic habitats and seek insect prey in fields and forest clearings brings them into close contact with humans, even in urban settings. The larvae of dragonflies are strictly aquatic and spend months and, in some species and habitats, years in freshwater prior to emerging as adults to fly, hunt, and reproduce.

As can be seen in Figure 7.9, dragonflies exhibit partial metamorphosis (technically hemimetabolism), that is, although the larval forms lack fully developed wings and adult gonadal characteristics, they possess many other features in common with the adults including jointed legs and easily recognized divisions of the body into head, thorax, and abdomen.

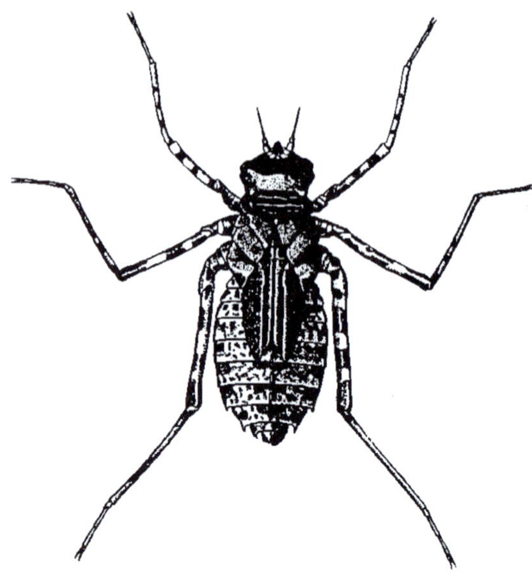

Fig. 7.9. A dragonfly larva. (Adapted from Borror et al., 1981.)

The cuticle of dragonfly larvae resembles that of other aquatic insects and therefore has a finite permeability to water. As such, dragonflies are subjected to a constant influx of water across the cuticle because of the osmotic gradient that exists between the external medium and their sodium-rich and chloride-rich hemolymph.

Dragonfly larvae are predaceous. They feed on any other animal they can catch and subdue, including other insects, crustaceans, annelids, and fish. From their animal prey, the larvae obtain not only nutrients but also valuable ions, particularly sodium, chloride, potassium, and calcium. The nutritional intake of ions unquestionably provides a valuable source of these ions to the larvae, sparing energy that would otherwise have to be expended on ion uptake from the medium. However, the larvae often go for days and weeks between fortuitous hunting outcomes, and the Malpighian tubules must produce dilute urine throughout this period to remove water that has entered through osmosis. During this period, the insects require a mechanism for extrarenal ion uptake.

In dragonfly larvae, the site of this uptake is the rectum. We have already mentioned that in insects urine is produced in the Malpighian tubules and modified through ion transport in the rectum. This occurs in dragonfly larvae as well. Urine is excreted through the anus once the ions have been retrieved to the greatest extent possible. Dragonfly larvae are also capable, however, of taking water back up through the anus into the rectal lumen. The anus, therefore, serves as a two-way valve for water entering and exiting. The insects use this water for three distinct purposes. Firstly, the lining of the rectal epithelium contains a densely branching network of tracheae. These serve as the site for gas exchange between water in the rectal lumen and the gas that facilitates diffusion along the tracheae through the body. In most insects, the rectal fluid would be very low in oxygen indeed, but in dragonfly larvae, the capability of moving external water into and out of the rectum, provides well-oxygenated water to this epithelium. Secondly, the rectum, which already has the capacity to take up ions from the urine, also takes up vital ions, such as sodium, chloride, and potassium, from the external water flowing into and out of the rectum. Once again, the site of rapid fluid flow across the respiratory surfaces serves as a site of ion uptake. Finally, dragonfly larvae are capable of using powerful muscles in the rectum to expel water at high speed out of the rectum through the anus. This stream of water serves as an aquatic jet, propelling the larvae forward with surprising speed. This form of locomotion is useful for ambushing their aquatic prey, or for themselves escaping hungry predators.

7.7 Crayfish

Arthropods have been extraordinarily successful in invading freshwater, and not all of these arthropods are insects. The crustaceans also display surprising diversity in freshwater habitats. We can explore their osmotic processes in freshwater by examining the crustacean that has been the subject of the greatest amount of scientific attention, the crayfish.

Crayfish possess a relatively impermeable external integument. The surface of arthropods is covered with tough cuticle that consists of cross-linked proteins and carbohydrates. In decapods (the group of crustaceans possessing 10 legs to which crayfish belong), this cuticle is often made tougher, harder, and less permeable by the inclusion of minerals such as calcium carbonate in the cuticle. As is the case with fish, however, the Achilles heel in the crayfish is the respiratory epithelium which must of necessity be thin, highly vascularized, and exposed to a stream of rapidly moving freshwater. The respiratory surface in crayfish consists of the gills that reside in the thorax protected by an external layer of carapace (Fig. 7.10a). The gills themselves are highly branched and ramified, such that a large surface area can be brought into contact with water flowing through the respiratory gill chamber. As a result, crayfish achieve effective respiratory gas exchange, but are subjected to substantial osmotic uptake of water across the gill epithelium.

The excess water that enters the animal due to osmosis is excreted in the urine. In crayfish, urine is produced by the antennal gland (Fig. 7.10b). Primary urine is produced by filtration in the coelomosac. This fluid is isos-

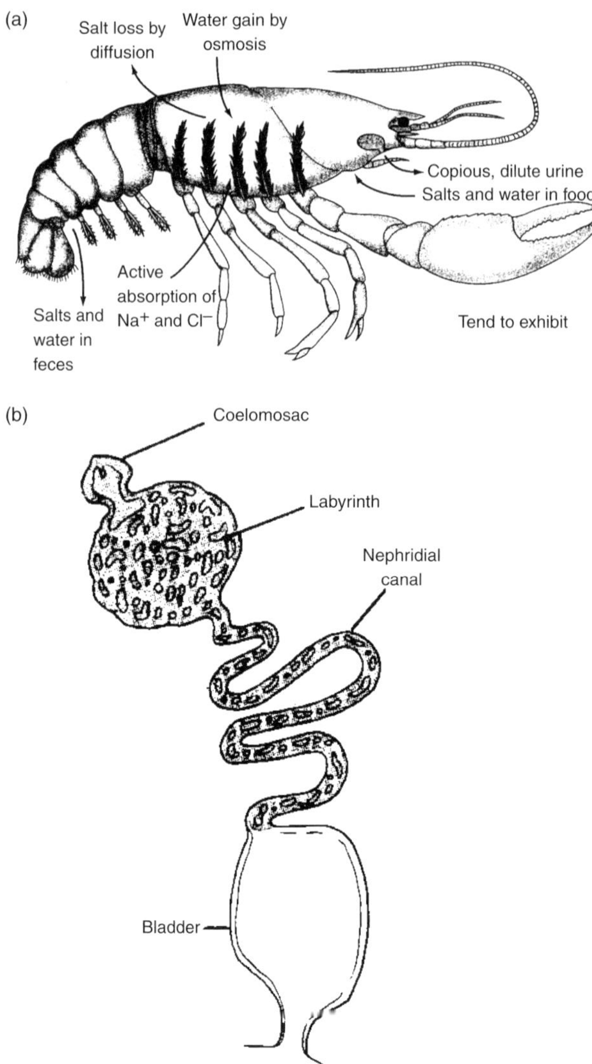

Fig. 7.10. (a) An illustration of a crayfish showing the location of the gills and the pathways of water and ion movements associated with osmoregulation. (Adapted from Hill et al., 2004.) (b) The antennal gland is the site of urine formation in crayfish. (Adapted from Randall et al., 1997.)

motic to the hemolymph and is rich in NaCl. As urine flows into the nephridial canal, ions are actively removed from the urine and returned to the hemolymph. It has been shown that up to 95% of the sodium in the primary urine is removed prior to excretion in freshwater decapods. The urine is thus substantially more dilute than the blood, but it is nonetheless much more concentrated than the external medium, necessitating an additional site for ion uptake.

As in fish, this capacity resides in the gills. The epithelial cells in the gills are capable of active Na^+ and Cl^- transport. This process restores lost ions to the hemolymph and maintains ionic homeostasis.

7.8 Mollusks

As stated in Chapter 6, the osmotic concentration of the blood and extracellular fluids in animals is isosmotic to that of the cells. In most animals residing in freshwater, the extracellular fluid is regulated in the range of 200–300 mOsm. As discussed in this chapter, this creates constant, ongoing physiological problems for freshwater animals due to the influx of water by osmosis. One obvious solution to this problem would be to lower the osmotic concentration of the blood, and therefore the tissues, to reduce this gradient. This solution runs up against the need of the animals to retain the required concentrations of proteins, ions, and nutrients in the cell cytoplasm. As a result, few freshwater animals possess the capacity to operate effectively with extracellular osmotic concentrations below 200 mOsm.

One group of animals does, however, attempt to take this strategy as far as they can. This group is the freshwater mussels. Bivalve mollusks have soft bodies that are contained inside the hard shells (termed valves) secreted by their mantle tissues. The shells consist of calcium carbonate secreted in a protein matrix. These shells are entirely impermeable to water and bivalve mussels face no osmotic challenges as long as the valves are closed tightly. Closing the valves, however, precludes any capacity to feed or respire and mussels are therefore forced to open their valves to carry out these necessary physiological functions.

Feeding in mussels involves the pumping of large quantities of the external water across the gills in order to capture and remove microscopic organisms and organic particles through filter feeding. Particles are captured from the stream of water passing over the gills through adherence to a mucous layer on the gill surface. This mucous is transported to the oral opening for ingestion. This form of feeding requires the juxtaposition of filtering tissues possessing a large surface area with rapidly flowing water.

Respiratory exchange between the mussel and its aquatic surroundings also occurs at the gills. The ultrastructure of mussel gills was described in Chapter 6 (Figs. 6.3 and 6.4). The gills by necessity have a very large surface area and a thin, delicate epithelium that separates the mussel's blood supply from the external medium. Blood, containing respiratory pigments flows through the gills on the internal surfaces of the gill filaments, separated by only a thin layer of tissues from the external medium flowing past. In addition, a number of other tissues in the mussel come into contact with the external medium, including the mantle and external surfaces of the gonads. As a result, the surface area over which osmotic flow can occur is very high in these animals.

Given the many tissues and large surface area exposed to the external medium in mussels, the energy that would be required to remove water entering the animal through osmotic flow is very great. To reduce this excretory and energetic burden, freshwater mussels have the lowest osmotic concentration in their body fluids of any animal measured (40 mOsm). This serves to reduce the flow of water substantially as the rate of water entry through osmosis at a given permeability is directly related to the difference in osmotic concentration. Nonetheless, the same inevitable rules of physics apply to mussels as to the other freshwater animals, they must produce dilute urine to rid the body of water and maintain volume regulation.

In advanced mollusks such as the mussel, urine is produced by filtration at the heart. A pericardial sac gathers a filtrate that is forced across the ventricular walls. This primary urine flows through tubules in which the ions are resorbed. As in the vertebrates and insects, therefore, the primary urine derives from the blood. This primary urine must be modified through the removal of ions and nutrients and the secretion of waste products. These transport processes occur in kidney tubules. As a result, the final urine secreted by the animal into the stream of water passing through the valve chambers is low in osmotic concentration and rich in nitrogenous and metabolic wastes.

As in all of the freshwater organisms, mussels must possess an extrarenal site of ion uptake. In mussels, this activity resides in the gills. It has been shown that the gills can actively absorb sodium and chloride from the external medium, replacing the ions lost from the organism by the production of a urine which is dilute but nonetheless more ion-rich than the medium.

7.9 Freshwater amoebae

Unquestionably, the animal facing the greatest surface to volume ratio in freshwater is the amoeba, an animal consisting of a single cell and surrounded not by an integument but rather by a single membrane composed of a lipid bilayer. Though many protozoa live in freshwater, amoebae have gained special attention because they do not possess a test or shell to reduce osmotic permeability. The form of locomotion engaged in by amoebae (amoeboid motion) requires that the entire membrane be flexible and motile.

The osmotic concentration of the cytoplasm of amoebae is similar to that of other animals. It follows that the cytoplasm must take up water by osmosis. Given that the animals consist of a single cell, how is urine produced and diluted to regulate cytoplasmic volume and osmotic concentration? As amoebae consist of a single cell, the term urine would be misleading so let us call it the excretory fluid.

Excretory fluid in amoebae is accumulated in a contractile vacuole (Fig. 7.11). If one observes an amoeba under the microscope one notices a series of cyclical events. Small vacuoles appear in the cytoplasm and migrate toward a central region. Here they accumulate and fuse, forming an even larger vacuole filled with fluid. Periodically, this vacuole contracts, expelling fluid through the cell membrane and into the external medium. Physiologists have sampled this fluid as it accumulates in the contractile vacuole prior to expulsion and determined that the fluid is very dilute, having much lower ionic concentrations than the cytoplasm of the amoeba. Because we have no evidence in any organism for active water transport, researchers have concluded that fluid must accumulate initially through ion transport into the small, forming vacuoles. As these proceed toward the contractile vacuole, ions are removed from the vacuoles, thereby forming a dilute fluid in the vacuoles. This fluid is accumulated in the contractile vacuole and is expelled as a dilute excretory fluid. The rate of fluid excretion is linked to the rate of fluid entry into the cell. This has been demonstrated by varying the osmotic concentration of the fluid around the amoebae. In more concentrated media, the contractile vacuole fills and empties less frequently than in dilute media.

Although the fluid expelled from the contractile vacuole is more dilute than the cytoplasm, it is more concentrated than the very dilute media in which amoebae can live. It follows that the amoebae must be able to take up ions directly from the medium, probably by means of active ion trans-

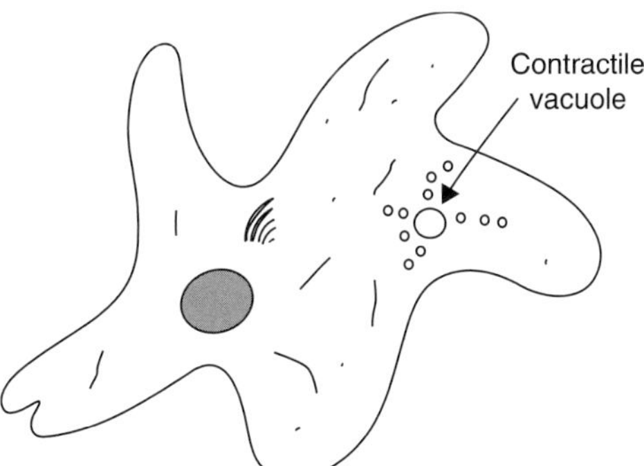

Fig. 7.11. An illustration of an amoeba showing the location of the contractile vacuole within it.

porting moieties in the plasma membrane. The difficulty of working with organisms as small as amoebae has made the unequivocal demonstration of this process difficult. Nonetheless, the maintenance of ion concentrations in the cytoplasm at levels well above the external medium is direct evidence of active processes for ion uptake.

7.10 Summary

The above examples illustrate that all animals in freshwater maintain body fluids with osmotic and ionic concentrations above those of the freshwater medium. Unless the integument is extremely impermeable (reptiles, birds, and mammals), this leads to the osmotic uptake of water into the body. All freshwater animals produce an excretory fluid that is more dilute than the blood, thereby restoring body volume and ridding the body of water obtained from the medium. All freshwater animals require some site of extrarenal ion uptake. These sites are quite diverse in different animal groups, existing on various portions of the external integument or the gut (Table 7.2).

Suggested additional readings

Borror, D.J., D.M. DeLong & C.A. Triplehorn (1981) *An Introduction to the Study of Insects*. Saunders College Publishing, Philadelphia, PA.

Edmunds, G.F., S.L. Jensen & L. Berner (1976) *The Mayflies of North and Central America*. University of Minnesota Press, Minneapolis, MN.

Evans, D.H., P.M. Piermarini & K.P. Choe (2005) The multifunctional fish gill: dominant site of gas exchange, osmoregulation, acid-base regulation and excretion of nitrogenous waste. *Physiol. Rev.* 85:97–177.

Hill, R.W., G.A. Wyse & M. Anderson (2004) *Animal Physiology*. Sinauer Associates, Sunderland, MA.

Patrick, M.L., K. Aimanova, H.R. Sanders & S.S. Gill (2006) P-type Na^+/K^+ ATPase and V-type H^+-ATPase expression patterns in the osmoregulatory organs of larval and adult mosquitoes, *Aedes aegypti*. *J. Exp. Biol.* 209:4638–4651.

Randall, D., Burggren, W. & French, K. (1997) *Eckert Animal Physiology*. W.H. Freeman & Co., New York.

Wood, C.M. & T.J. Shuttleworth, eds. (1995) *Cellular and Molecular Approaches to Fish Ionic Regulation*. Academic Press, San Diego, CA.

8 Terrestrial animals

8.1 Introduction

In the instance of aquatic animals, we have discussed at length the gradients for the activity of water between the external medium and the body fluids. Terrestrial animals live in an environment in which the external medium is not aquatic, but rather consists of land and air. How then are we going to understand the osmotic forces that apply under these circumstances and the mechanisms that the animals might use to regulate the osmotic concentration of body fluids? To begin this process, let us return to some basic principles of physical chemistry related to the activity of water in aqueous media and air.

Consider the situation shown in Figure 8.1, in which we have a beaker containing distilled water. The top part of the beaker contains air and the beaker is sealed by an impermeable lid. Eventually, an equilibrium will be set up between the water in the beaker and the water vapor in the air above it. As the water is pure distilled water, the osmotic concentration of the water is 0 mOsm.

The amount of water vapor in the air above the water can be expressed in terms of relative humidity (% RH). A relative humidity of 100% means that the air is fully saturated with water vapor, that is, it contains all the water vapor it can at that temperature. When air comes into equilibrium with the distilled water, the relative humidity will be 100% RH. If, just through random molecular movements, the amount of water vapor in the air should drop a bit, water would evaporate from the liquid water surface, returning the two phases to equilibrium. If excess water vapor should accumulate in the air, then this would result in water precipitating

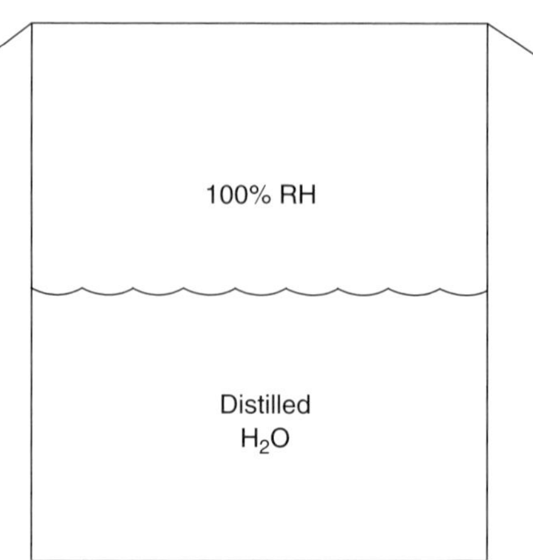

Fig. 8.1. At equilibrium, the air above pure water will be at 100% relative humidity (100% RH).

from the air, returning it to the liquid phase and restoring the equilibrium condition.

Another way of expressing this is that the water in its gaseous phase also has an activity and, in an equilibrium condition, the activity of water in the liquid and aqueous phases are equal. If we lower the activity of water in the aqueous medium, we would lower it in the air as well. One way of lowering the activity of water is to add solutes. Let us consider a situation similar to Figure 8.1 but with seawater in the beaker as opposed to distilled water (Fig. 8.2). Seawater has an osmotic concentration of about 1000 mOsm. It would come into equilibrium at 25°C with a relative humidity of less than 100%, actually about 98% RH. In other words, the air would contain less water vapor because the driving force for water entry into the air is the activity gradient for water between the air and the aqueous medium. As the activity of water has been lowered in the aqueous phase, it must also be lowered in the air at equilibrium.

If we add further solutes to the seawater, the relative humidity in the air with which it is in equilibrium would decline further. This principle is used in vapor pressure osmometers. They measure the osmotic concentration of fluids by measuring the vapor pressure of water in the air that has come into equilibrium with the fluid.

The activity of water is also affected by temperature. As temperature goes up, the activity of water in an aqueous solution also rises. As a result, air that is in equilibrium with that solution can also hold more water in vapor phase. If we return to the conditions shown in Figure 8.1, the amount of water vapor in the air depends on temperature. Table 8.1 shows the vapor pressure generated by water at various temperatures.

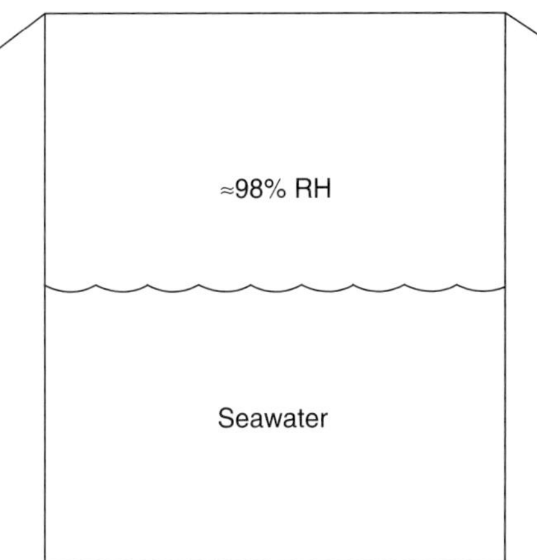

Fig. 8.2. At equilibrium and at 20°C, the air above seawater will be at about 98% RH.

Table 8.1. The relationship between temperature and the vapor pressure of pure water at that temperature

Temperature (°C)	*Equilibrium vapor pressure (kPa)*
0	0.6
10	1.2
20	2.3
30	4.3
40	7.3
50	12.4
60	19.8
70	31.0
80	47.2
90	69.8
100	101

Note that the vapor pressure of water goes up with temperature. At 100°C, the vapor pressure of water match with the atmospheric pressure at sea level (101 kPa).

The effects of temperature on the capacity of air to hold water vapor have implications for animals with regard to water loss. If a terrestrial animal is exposed to a cool wind in which the air has a water content below saturation, then the animal will be exposed to dry conditions as water evaporates from its surface into the passing air. If the animal is exposed to a warm wind with the same water content, then the animal would lose water more rapidly to the passing air because the activity gradient for water between the skin and the air would be larger.

This has implications for the effects of body temperature as well. As the activity of water increases with increasing temperature, water evaporates more rapidly at high body temperature than at low body temperature.

Relative humidity is an expression of the amount of water vapor in air relative to the amount that would be present at saturation. The amount of water vapor in the air at saturation increases with increasing temperature. Therefore, air with a humidity of 80% RH at 10°C contains less water vapor than does air at 80% RH at 25°C. Relative humidity is a useful term for many purposes, but because of the effects of temperature, it is not a useful measure of the driving force for the passive movement of water into the vapor phase. For those quantitative purposes, animal physiologists tend to employ the partial pressure of water vapor, expressed in kilopascals (kPa).

8.2 The terrestrial environment can be associated with enormous gradients for the activity of water

While discussing the capacity of aquatic animals to survive in saline waters, we mentioned the champions of osmotic regulation, the brine flies and brine shrimp (see Chapter 6). These highly saline-tolerant animals can survive in media as concentrated as 6–10 times the concentration of seawater. Saline medium of this concentration is in equilibrium with air at about 88% RH at 20°C. As you are well aware, a day with 88% RH feels fairly humid. In most climates, relative humidity frequently falls below this value. This means that the osmotic gradient, that is, the gradient for the activity of water between the blood of terrestrial animals and the air, can potentially be much larger than the gradient experienced by the most saline-tolerant aquatic animals. Any time the air contains less than 98% RH at 20°C, that is, almost any time when it is not raining, the gradient for the activity of water across your skin is greater than that experienced by any marine animal. We can see, therefore, that the terrestrial environment presents an enormous challenge for animals with regard to water regulation and retention. The remainder of this chapter will deal with the many mechanisms employed by terrestrial animals to deal with this challenge.

8.3 Osmoregulatory strategies among terrestrial animals

The above discussion established that terrestrial animals face the largest gradients for the activity of water across their integuments of any animal.

As a result, the terrestrial environment is a very challenging one from an osmoregulatory point of view. Only three phyla of animals have managed to fully exploit the terrestrial environment: the vertebrates, the arthropods, and a few mollusk species.

A key adaptation in most terrestrial species is the mechanism for reducing the permeability of the integument with regard to water. The mechanisms to achieve this vary with the animal group, but given the huge gradient for the activity of water that exists in the terrestrial environments, the permeability exhibited by most aquatic organisms would lead to death in the terrestrial arena. A particular challenge in this regard is posed by the respiratory surfaces, which must remain highly permeable to oxygen and carbon dioxide, while minimizing the loss of water.

Acquisition of water is another key feature of terrestrial life. The water may be obtained as free water, or in the food upon which the animal feeds, but access to water is critical for all terrestrial animals. Some arthropods can obtain water in the form of water vapor and we will discuss this process in more detail below.

Finally, those animals inhabiting the most challenging terrestrial environments often possess specialized organs for producing and excreting a fluid more concentrated than the blood. By producing a hyperosmotic secretion, these animals raise the activity of water in their body fluids, thereby protecting the proteins and membranes in their cells from potential denaturation due to high osmotic concentrations.

8.4 Terrestrial vertebrates

8.4.1 Amphibians

Most amphibians are restricted to aquatic or very moist terrestrial environments. Salamanders, for example, have highly permeable skins and can survive in terrestrial environments only if the activity of water in the air is very high. Salamanders tend to remain in moist or underground sites during the day when relative humidity is low and venture out at night when relative humidity is high.

A few amphibia, especially frogs and toads, can survive in terrestrial environments. Some, such as tree frogs, reside in sites where the relative humidity can be low and water is at a premium. Several toad species actually occur in deserts and survive for years in sites where rain is very sparse.

The osmotic concentration of the blood of all vertebrates is regulated at a value around 250–300 mOsm. None of the amphibians is capable of producing a concentrated urine in the kidneys, or for that matter any kind of fluid that is hyperosmotic to the blood. Therefore, the osmoregulatory strategy for the amphibians consists simply of assuring access to water and reducing water loss to a minimum.

Amphibian can obtain water from their food although, because most amphibians are carnivorous, this rarely provides a highly dilute source of water. Frogs and toads rarely drink water. They certainly need to obtain water from the environment and particularly must be able to exploit not only standing water but also water in moist habitats. They do so by absorbing water directly across the skin, rather than by drinking. Many frogs and toads possess water-absorbing patches of skin on their ventral surfaces in their pelvic regions (Fig. 8.3). The osmotic permeability of the patches can be adjusted by the insertion or removal of aquaporins in the epithelial cells. If the frogs sense that the medium is moist and dilute (e.g., on a wet leaf, or a spot of muddy soil), then aquaporins are inserted into the epithelial cell membranes thereby increasing their permeability to water. Water flows down its activity gradient into the amphibian's cells and into

Fig. 8.3. A frog sitting on a pane of glass, viewed from below. The pelvic patch is visible as a dark region between the hind legs and in the pelvic region of the belly. (Adapted from Hillyard *et al.*, 1998.)

the bloodstream. The water patches are highly vascularized, facilitating the distribution of water from the skin into the body. Frogs and toads cannot actively remove water from the environment. The movement of water is instead always passive and is generally not associated with active ion transport.

If water-stressed amphibians can produce a urine that is close to that of the blood in osmotic concentration, then the necessity for eliminating metabolic wastes occurs with a minimum water loss. Many amphibians store urine in the bladder rather than excreting it. At times when the animals are well hydrated, they produce dilute urine and store it. Subsequently, when exposed to dry conditions, the animals can retrieve water from the urine in the bladder as they dehydrate. The bladder is capable of recycling ions from the urine as well. This lowers the osmotic concentration of the urine and then water flows osmotically through the bladder wall back into the blood. The urine stored in the bladder can be an important source of water for animals facing long periods without water.

Toads tend to have more highly keratinized skin than do frogs. As a result, they generally have lower rates of water loss across the skin than do frogs. This is an important adaptation in toads that accounts for their distribution in drier habitats than frogs.

Some frogs, in turn, are even more impermeable than toads. They have an additional mechanism for reducing water loss through the integument, namely, the secretion of waxes by glands in the skin. They then spread these water-resistant hydrocarbons over the skin using their limbs. Coating themselves with this layer of wax molecules reduces the rate of water loss substantially. Frogs using waxes to reduce their rates of integumental water loss are usually found in the tropics where higher temperatures make the application of waxes easier and the retention of water more critical.

8.4.2 Reptiles

The osmotic gradient across the integument, that is, between the blood and the external air, is identical in all of the terrestrial vertebrates. Amphibians, reptiles, birds, and mammals all have blood concentrations of around 250–300 mOsm. Therefore, at all relative humidities below about 99%, the gradient for the activity of water favors water loss across the skin. Reptiles, birds, and mammals all possess keratinized skin, an important adaptive feature for terrestrial life. The epidermal cells of these animals are filled with a large filamentous protein termed keratin. As the cells age, they migrate to the surface of the integument and flatten. The cells eventually die and are pressed to the surface by new epidermal cells proliferating under them.

This causes the surface of the skin to be covered by thick layers of dead, flattened cells that are filled with tough bundles of keratin. These cells provide a formidable barrier to water loss because of the multiple layers of lipid-containing membranes they contain. The keratin serves to make the skin much tougher and resistant to cutting and abrasion. Despite their thick and heavily keratinized skin, reptiles continue to lose water across their integument, albeit at a very low rate compared to amphibians.

An additional source of water loss for reptiles is the respiratory tract. Consider a reptile which finds itself in a situation where the relative humidity is 80% at 20°C. The animal breathes in a lungful of air that upon entering the lungs comes into equilibrium with the moisture on the lung surface. This fluid is, in turn, in equilibrium with the extracellular fluid around the lung tissue. To saturate the air in the lungs, or at least bring it to 99% RH, water is removed from the body fluids. As only water but not solutes are lost in this process, this process increases the osmotic concentration of the body fluids. When the air is exhaled, the water is lost from the body in the form of water vapor. Water is lost with every breath. The higher the metabolism, the more rapidly the reptile breathes, and the faster the water is lost.

Therefore, due to cutaneous and respiratory water loss, terrestrial reptiles almost always experience some water loss to a terrestrial environment that has a lower activity of water than the body fluids. The obvious and indeed most frequently employed solution to this problem is the ingestion of fluids more dilute than the body fluids. Many reptiles will drink fresh water and this serves to dilute the body fluids and replace the water lost through evaporation. Similarly, many reptiles feed on leaves or fruits, and the tissues of plants almost always have a higher activity of water than the body fluids of vertebrates.

Some reptiles, for example, desert animals, may not have access to fluid more dilute than their body fluids for months at a time. If, similar to some snakes and lizards, they feed on mammals and birds, their food contains fluids isosmotic to their own. Other reptiles feed on insects that also have body fluids with concentrations around 250–350 mOsm.

The kidneys of reptiles, similar to those of fish and amphibians, are unable to produce urine more concentrated than the blood. They lack the loops of Henle that birds and mammals possess, and thus cannot produce hyperosmotic urine. In reptiles, the urine is produced by filtration in the glomerulus. This process is a simple form of physical filtration in which large molecules and cells are removed, but the fluid passes through relatively unchanged with regard to small solutes. The resulting primary urine is identical to the blood in osmotic concentration. Metabolically important solutes are removed as the urine passes through the proximal tubules. As the proximal tubules are osmotically highly permeable, the urine remains isosmotic with the blood. In the distal tubules, solutes can be removed but not added. The urine, therefore, can be made more dilute than the blood but not more concentrated.

Similar to marine iguanas (see Chapter 6), some terrestrial species of lizards can secrete a concentrated fluid from nasal glands. This secretion can be rich in sodium chloride, but in some desert species that feed on plants, the salt glands can secrete a hyperosmotic fluid rich in potassium and chloride, the major ions that they ingest.

8.4.3 Birds

Flying birds are quite mobile in the terrestrial environment and take advantage of this mobility to find sources of water as required. Similar to reptiles, birds have a relatively impermeable skin. Unlike reptiles that frequently have scales to increase the toughness and impermeability of the skin, birds often possess feathers. Birds also lose water through the respiratory tract to the dry air that passes through the lungs. Birds are warm-blooded (homeothermic) and therefore must regulate their body temperature. In a warm environment, birds can get overheated and must dump heat to the environment. They do this by depressing their feathers to reduce insulation, but also by panting as a means of increasing water evaporation from the mouth and tracheal surfaces. This activity causes the loss of water, exacerbating the problem of osmotic regulation for a terrestrial animal.

Urine in birds is produced in the kidneys. The urine exiting the kidneys passes into the cloaca in birds. If the urine is more dilute than the blood, water can be resorbed osmotically in the cloaca and returned to the blood. This serves to conserve water in the body even if the kidneys are not producing concentrated urine. The kidneys of birds, unlike those of reptiles, however, are able to produce urine more concentrated than the blood. The production of concentrated urine serves to maintain dilute body fluids and rid the body of salts obtained in the food. When the kidneys are producing concentrated urine, this urine can pass through the cloaca and remain hyperosmotic until excreted. Although the cloaca has the potential for the uptake of ions and water required by the animal, it is can also be made impermeable to allow concentrated urine to pass through.

Similar to marine birds (see Fig. 6.3), many species of terrestrial birds can produce a hyperosmotic secretion in nasal salt glands. This fluid serves to remove ions from the blood and to reduce the osmotic concentration of the body fluids. In all birds examined to date the nasal secretions are rich in sodium and chloride.

The capacity of birds to survive in dry terrestrial habitats is, therefore, due to a suite of behavioral and physiological adaptations. Birds seek out water and succulent food to provide water in terrestrial environments. The kidneys can produce concentrated urine and the cloaca provides an additional opportunity to regulate ionic and osmotic concentrations in the blood. Finally, in many birds the nasal salt glands serve to remove excess NaCl under conditions of high osmotic stress.

8.4.4 Mammals

Similar to birds and reptiles, mammals have a relatively impermeable skin compared to amphibians. In many mammals, fur also protects the skin, reducing the heat in sunlight, and reducing the evaporation by minimizing air movements adjacent to the skin. Nonetheless, some water loss occurs across the integument in mammals.

Some mammals, including humans and horses, use dermal sweat glands to place water (sweat) on the skin's surface. This process serves to reduce body temperature in warm environments or during exercise, but it can place substantial osmotic stress on the body. The primary secretion which is produced deep in the sweat glands is isosmotic to the blood. As the fluid flows toward the skin surface inside the sweat duct, salts (principally sodium and chloride) are removed to produce a hypo-osmotic fluid by the time it reaches the skin surface. When secreted slowly, such as would occur if a human was sitting in a warm room, the sweat is dilute. If the sweat is produced more rapidly, for example, during exercise on a warm day, the salt content is much higher because the fluid flows through the ducts faster than the removal of salts. Under any circumstances, sweating leads to the loss of blood volume, body water, and salts.

Mammals also experience respiratory water loss. As pointed out in the section above on reptiles, inhalation of dry air and exhalation of hydrated air leads to water loss. The situation in warm-blooded animals is a little more complicated due to differences in temperature between the environment and the animal's body. Let us take the example of a mammal breathing air with a relative humidity of 60% RH and a temperature of 20°C. When the mammal breathes in, the air passes across the nasal passages, evaporating water from these surfaces. This serves to hydrate the air. It also cools the nasal passages as the air is cooler than the body temperature and evaporative cooling is also occurring. If the air is not fully hydrated in the nasal passage, it will take up additional water from the tracheal surfaces, so that the air is fully saturated with water vapor when in the lungs. The air in the lungs also comes into temperature equilibrium with the body. As pointed out in the beginning of this chapter, air at 36°C contains more water at saturation than does air at 20°C. Therefore, by warming the air and saturating it with water, the mammal has lost water to the air. At this point, the mammal will exhale. As the air is expelled it passes out of the nose. Here it comes in contact with the nasal passages which were cooled during inhalation. The nasal passages cool the air passing by. As the exhaled air is saturated with water vapor, any cooling leads to condensation. Water condenses on the nasal epithelia as the air passes by. The nasal epithelia are osmotically quite permeable and this water is osmotically absorbed into the underlying body fluids. Many mammals have elaborate nasal turbinates in the nasal passages which increase the surface area available for contact with inhaled and exhaled air. These passages play an important role in reducing respiratory water loss.

In mammals, the most important organs with regard to osmotic regulation are the kidneys. Urine formation in all of the vertebrates (with the exception of aglomerular fishes, see Chapter 6) occurs in the glomeruli. Here, blood pressure is used to filter the blood through a permeable barrier composed of the capillary wall, basal lamina, and podocytes. The filtered primary urine is osmotically and ionically identical to the blood plasma with the exception that blood cells and large proteins have been filtered out. This primary urine flows into the proximal tubules in which ions and metabolically valuable solutes are actively removed and returned to the circulatory system. The removal of these abundant small solutes, such as amino acids, sugars, and salts, substantially reduces the quantity of osmotically active solutes in the primary urine. The proximal tubules are osmotically quite permeable so that the solutes are returned to the blood, water follows. As a result, the volume of the urine is substantially reduced (by about 90%) but the osmotic concentration does not change at all.

In the distal tubules, active uptake of sodium and chloride occurs. In the distal tubule, however, the epithelium is less osmotically permeable and the removal of ions results in a dilute urine. In most vertebrates, it is this activity that is responsible for the production of a dilute urine.

In all mammals, the urine passes from the proximal tubules to the distal tubules through a thin connecting segment. The process of producing a concentrated urine involves the activities of this narrow tubule segment connecting the proximal and distal tubules, termed the loop of Henle (Fig. 8.4). In some nephrons the connecting section is short and the fluid flows almost directly into the distal tubule. In other nephrons and particularly in those species capable of producing a concentrated urine, the interconnecting thin sections are longer. The loop of Henle is divided into two morphologically and functionally distinct regions. Urine flowing from the proximal tubule enters the descending limb of the loop which extends into the middle (medulla) of the kidney. Here the tubule makes a sharp u-turn and the urine flows in the ascending limb, returning to the cortex. The urine then exits the ascending limb and flows into the distal tubule.

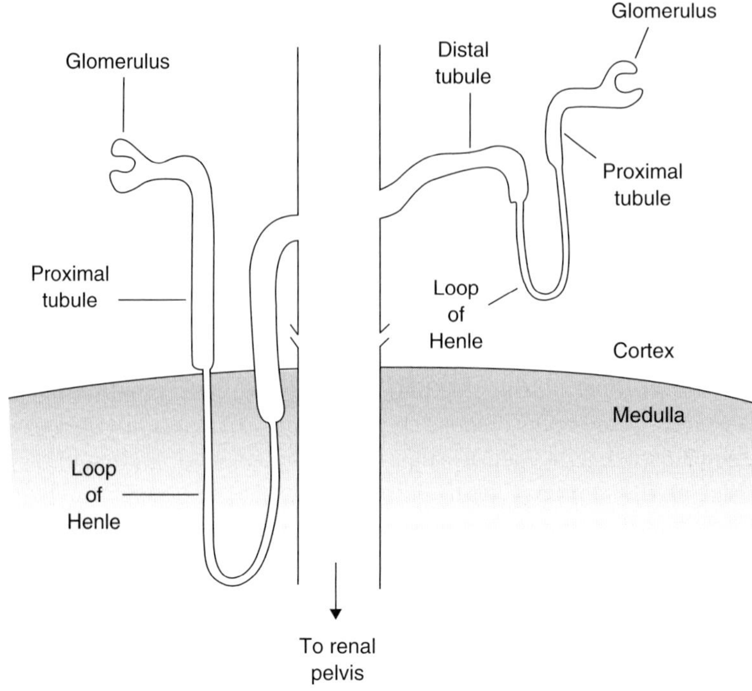

Fig. 8.4. The kidney tubules of mammals have glomeruli in which the primary urine is formed by filtration. The primary urine flows sequentially into the proximal tubule, the loop of Henle, the distal tubule, and then into the collecting duct. The outer region of the kidney is termed the cortex, while the more central region is termed the medulla. Some kidney tubules (shown on the right) have short loops of Henle that do not descend into the medulla. These are termed cortical tubules. Others, termed juxtamedullary tubules, (shown on the left) have long loops that descend deep into the medullary region. The latter tubules are responsible for maintaining the highly concentrated interstitial fluids in the medulla of those mammals able to produce hyperosmotic urine.

The medulla of the kidney in many mammals has an osmotic concentration that is much higher than that of the blood. As we have stated, the blood of mammals is around 280–350 mOsm. The medulla of the kidney can have an osmotic concentration as high as 1000–1800 mOsm depending on the species. Concentration of the fluid in the medulla occurs in the following manner. As the urine flows down into the medulla in the descending limb, water is osmotically withdrawn from the urine into the medullary extracellular spaces. The descending limb is permeable to water but not very permeable to ions so that the urine comes into equilibrium with the highly concentrated medullary region, which is enriched in ions. The urine flows around the loop and begins its return to the cortical region via the ascending limb. In this region of the nephron, sodium and chloride are actively removed from the urine and transported into the medullary extracellular spaces. It is this active transport of ions that forms a highly concentrated urine that produces the elevated osmotic concentration in the medulla. The removal of ions from the urine flowing through the ascending limb is so effective that by the time the urine reaches the distal tubules it is actually hypo-osmotic to the blood.

As the urine descending into the medulla in the descending limb loses water and the urine flowing in the ascending limb loses salts, one might expect these two processes to cancel each other. It is worth remembering, however, that the urine flowing into the limb from the proximal tubule is isosmotic to the blood. The urine flowing out of the ascending limb into the distal tubule is hypo-osmotic. This means that ions were removed in the medulla and a disproportionately small amount of water followed. It is this accumulation of salts without proportional water following that forms the concentrated extracellular fluid in the medulla. You will remember from Chapter 4 that the cells in multicellular animals are always isosmotic to their surrounding extracellular fluid. This is true of the mammalian kidney as well. The cells in the medulla, and this includes kidney tubule cells, endothelial cells lining the circulatory system, and blood cells, all must come into equilibrium with this highly concentrated fluid. The blood cells pass rapidly through these areas, but they do experience a substantial increase in osmotic concentration and loss of volume during this short passage. The kidney and endothelial cells must remain in this region for the lifetime of the animal. They are able to survive in this highly concentrated environment by accumulating compatible solutes, particularly amino acids and sorbitol. In this respect, the cells in the renal medulla resemble the cells of marine organisms which must also withstand similar osmotic concentrations.

The length of the loop of Henle determines the capacity of the kidney to produce concentrated urine. Aquatic mammals, such as the beaver, that have no need to produce concentrated urine, lack a long loop of Henle. Camels, in contrast, produce a very concentrated urine and their loops of Henle are very long and descend deep into the medulla of the kidney. Humans can produce a concentrated urine as well, but not as concentrated as a camel's. The length of the loop of Henle in humans is intermediate. All mammals, including humans, have nephrons that vary with regard to the length of the loop of Henle. Some loops are short, transporting the urine almost directly from the proximal tubules to the distal tubules. Others are long and dip deeply into the medulla. It is these latter tubules that maintain the high osmotic concentration in the medulla. The flow of urine through one type of nephron or another is controlled by variations in the flow of blood through the glomeruli producing urine for that nephron.

In our description of mammalian kidney function to this point, we have produced an isosmotic urine in the glomerulus, removed most of the metabolically important nutrients in the proximal tubules, removed ions to concentrate the medullary extracellular space, and removed ions in the distal tubule to produce dilute urine. The production of dilute urine does not, however, solve the osmotic regulatory problems of terrestrial animals. We need to identify some mechanisms for concentrating the urine.

The urine flowing down numerous distal tubules (which you will remember as hypo-osmotic to the blood) is brought together in the collecting ducts. The urine flowing in the collecting ducts passes back down through the medulla of the kidney. If the epithelia of the collecting ducts are impermeable to water, the urine remains dilute and can pass through the ureter to the bladder for excretion. Under these conditions a dilute urine is produced and eliminated. In contrast, if the epithelia of the collecting ducts become permeable to water, the urine passing through the collecting ducts can loose water to the surrounding medullary spaces, and can in fact become almost as concentrated as the fluid in the spaces. In humans, this means that the urine can approach 1200 mOsm in concentration, markedly hyperosmotic to the blood which is about 300 mOsm.

Control of the osmotic permeability of the collecting ducts is, therefore, the mechanism by which the concentration of urine is regulated. How can the permeability of an epithelium be rapidly modified and controlled? The answer lies with the location and function of aquaporins. When they lack aquaporin molecules, the apical membranes of the collecting ducts are highly impermeable to water. Under these conditions dilute urine flows through the collecting ducts and down into the bladder. The cells of mammalian collecting ducts contain vesicles with aquaporin molecules embedded in their membranes. As long as these aquaporin molecules are contained in the vesicle, they have no effect on the permeability of the apical membranes. If the blood becomes excessively concentrated, however, antidiuretic hormone (ADH) is released and this promotes the fusion of the vesicles with the apical membranes. This serves to insert the aquaporin molecules into the apical membrane greatly increasing its osmotic permeability (see Chapter 11 for more details about the hormonal regulation of this system). The basal membrane also is already permeable, so the change in the apical membrane determines osmotic flow across the entire epithelium. Under the influence of the ADH, therefore, water is drawn from the urine in the collecting ducts using the osmotic gradient established between the medullary extracellular spaces and the urine. The osmotic gradient, you will remember, is produced through the activity of the ion pumps in the ascending limb of the loop of Henle.

The production of urine of variable osmotic concentrations is vital for the survival of most mammals in the terrestrial environment. A review of the

Fig. 8.5. The kangaroo rat, *Dipodomys spectabilis*. (Adapted from Schmidt-Nielsen, 1994.)

mechanisms available to terrestrial mammals can be obtained by examining the processes used by a small desert rodent, *Dipodomys*, the kangaroo rat (Fig. 8.5). The desert in which this rodent is found provides it with almost no opportunities for drinking free-standing water. The animals obtain water from the metabolism of food. They live in burrows in which they store the seeds and vegetable matter on which they feed. Water lost by evaporation from their bodies can in part be absorbed by their food and be available upon ingestion. *Dipodomys* has extensive nasal turbinates and this aids considerably in reducing respiratory water loss. Finally, the rodent has long loops of Henle and can produce concentrated urine. Behavioral adaptations are quite important for the survival of these mammals as well. Kangaroo rats will feed on succulent food when it is available, and their nocturnal habits allow them to avoid the hot daylight hours in their burrows and roam about the desert floor only during the cooler nighttime hours.

8.4.5 Desert locusts

The terrestrial animals discussed above are all vertebrates. The organs they use to produce urine are the kidneys. Any additional sites of hyperosmotic fluid secretion (if any) are located in the head. Insects evolved from marine arthropod ancestors that moved out to land entirely independent of the vertebrates. The body plan of the insects is quite different from the body plan of vertebrates, and their evolutionary history is also distinct. Although insects face the same physiological constraints to terrestrial life (infrequent access to water, low humidity, low-sodium foods), the organs they use and the physiological mechanisms they employ are quite distinct from those of the vertebrates.

We will explore the processes used by the desert locust, *Schistocerca gregaria*. This insect is responsible for the locust plagues which occur in Africa and in the Middle East. Owing to its extreme economic importance, this insect has been studied in considerable detail. It can survive even in the most inhospitable deserts on earth. Not surprisingly, therefore, the physiological mechanisms responsible for osmoregulation in a terrestrial setting are highly developed and quite evident.

Figure 8.6 illustrates the internal organs of a desert locust. Food macerated by the mandibles is swallowed and then passes through the foregut into the midgut. Here the food is digested, through the actions of digestive enzymes, and absorbed. The cells lining the midgut contain transport proteins that drive the uptake of ions and nutrients contained in the meal. These nutrients include amino acids, fatty acids, and sugars produced by the enzymatic cleavage of larger molecules in the food. As a result of the uptake of these numerous, osmotically active solutes, the activity of water in the gut contents is substantially reduced. Water, therefore, enters the insect across the midgut wall, moving down its activity gradient.

When the ions, nutrients, and water cross the midgut wall, they enter the hemolymph. This fluid is the equivalent of blood in vertebrates. Hemolymph differs from vertebrate blood in that it lacks red blood cells containing hemoglobin and it circulates in a large open circulatory system lacking elaborate branching vessels.

Having obtained ions, nutrients, and water, the insects face the usual challenges of osmotic regulation, volume regulation, and waste excretion.

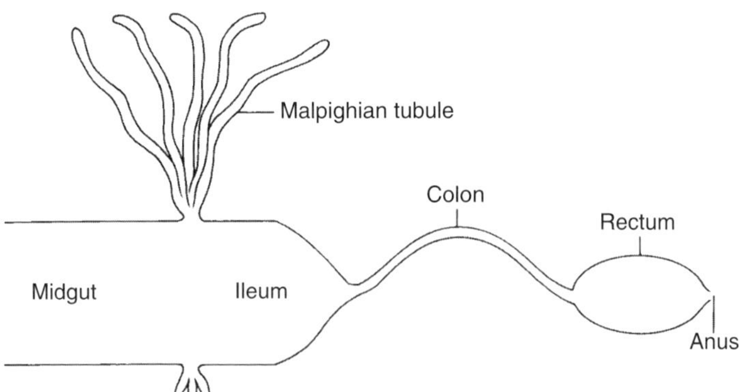

Fig. 8.6. A diagrammatic representation of the digestive and excretory organs of the desert locust. (Adapted from Bradley, 1985.)

Insects do not have kidneys. Instead, the site of urine formation is the Malpighian tubules. These tubules are narrow outpocketings of the gut that extend out into the hemolymph (Fig. 8.6). Their distal ends are closed and the proximal ends empty into the gut posterior to the midgut. The Malpighian tubules produce urine through the active transport of ions (see Chapter 9 for a more detailed description on ion transport processes in insects). Unlike vertebrate glomeruli that produce urine by ultrafiltration using hydrostatic pressure, insect Malpighian tubules actively transport ions across a permeable epithelium, allowing water to follow osmotically. The urine produced is isosmotic to the hemolymph and is rich in ions that are used to produce the osmotic gradient. In the case of the desert locust, these ions are principally potassium and chloride.

The fluid secreted by the Malpighian tubules serves to remove excess fluid and wastes from the hemolymph, but it has no effect on the osmotic concentration of the hemolymph as the fluid produced is isosmotic to the hemolymph. The fluid passes posteriorly in the gut, entering the ileum. Here active ion resorption, particularly of potassium and chloride, occurs. This activity serves to reduce the volume of the fluid passing down the gut. Valuable ions and water are retained and wastes are concentrated in the fluid remaining in the gut. The walls of the ileum are osmotically quite permeable so the fluid at this point remains isosmotic to the hemolymph.

This modified urine flows posteriorly into the rectum. The recta of terrestrial insects consist of a simple epithelium. This histological term means that the epithelium is one cell thick. Much of the epithelium is thin and distensible, which allows the rectum to swell when filled with urine and/or feces. Other parts of the epithelium are composed of thick cells (the rectal pad cells), which have highly folded apical and basal membranes and abundant mitochondria. The rectal pad cells can actively transport ions from the rectal lumen into the hemolymph. In this manner they can, if needed, retain the ions in the urine prior to excretion from the body through the anus. The epithelium of the rectum can be quite impermeable to water so the rectum is capable of resorbing ions faster than water can follow osmotically. In this manner, the rectum can produce dilute urine that is hypo-osmotic to the hemolymph. This would be the circumstance if the insects are eating abundant succulent foliage and thus loading the system with dilute fluids.

A common circumstance for an insect living in the desert, however, is the need to retain water and excrete ions, and to preserve the relatively dilute state of its body fluids in the face of the extreme heat and dryness of the surrounding air. The most critical need in these circumstances is the retention of water in the body through the production of concentrated urine. In terrestrial insects, these activities are also carried out in the rectal pad cells.

Figure 8.7 illustrates the morphology of the rectal pad cells. The apical surface of the cell, adjacent to the rectal lumen, has numerous, deep membrane infoldings. These membranes are the site of active ion transport from the rectal lumen into the cell cytoplasm. The ions absorbed are then transported into the intercellular spaces, forming there a fluid with very high osmotic concentration. The septate junctions between the cells can act as semipermeable membranes, permitting water to diffuse down its activity gradient from the rectal lumen into the fluid-filled intercellular clefts. As this water flows into the clefts, the fluid already in the clefts are forced in a basal direction through the intercellular spaces. In the more basal regions of clefts, ions are removed from the fluid and returned to the cell cytoplasm of the rectal cells. This ion transport activity removes ions from the fluid in a region in which the membrane is relatively impermeable to water.

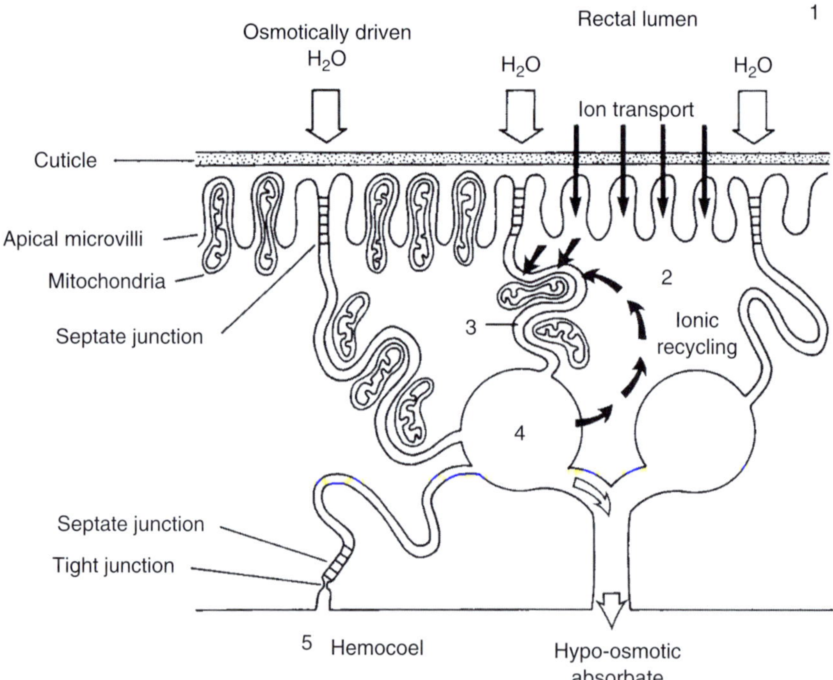

Fig. 8.7. A diagrammatic representation of the rectal pad cells of the desert locust. Active ion transport (shown with solid arrows) across the luminal membrane transports ions into the rectal cell and then into the intercellular clefts. Water follows (unfilled arrows) down its activity gradient. This concentrated fluid flows between the cells into the lower intercellular clefts. Here ions are transported back into the cells through membranes with a low osmotic permeability. The resulting dilute fluid flows into the hemolymph. The rectal cells are able to recycle the ions, thereby drawing water from the rectal lumen. (Adapted from Bradley, 1985.)

This serves to produce a dilute fluid because the major osmotically active solutes are removed and water does not follow. This dilute fluid continues to flow in basal direction and eventually passes into the hemolymph. The result of all of these activities is that water is drawn from the rectal lumen, is moved basally through the epithelium, ions are removed, and a dilute fluid is added to the hemolymph. This process retains water in the body and adds dilute fluid to the hemolymph, thereby counteracting the loss of water from the body resulting from evaporative and respiratory water loss.

The ions removed from the fluid in the intercellular clefts are transported back into the rectal cells. These ions are then recycled into the more apical regions of the cells, and used to draw more water from the rectal lumen. It has been shown that the rectum is capable of actively drawing water from the lumen using almost exclusively these recycled ions while drawing very few ions from the lumen. This allows the rectum to transport a hypo-osmotic fluid to the hemolymph while removing few if any ions from the rectal lumen. In this manner, the insects can produce concentrated excreta, retain water, and yet not load the body fluids with unneeded ions.

When the insects are well hydrated, for example, when feeding on abundant lush vegetation, the rectum can retain the needed ions and yet produce abundant dilute urine. During desiccating times, the rectum can remove water from the excreta while retaining the minimal number of ions required to serve physiological functions. The removal of water from the excreta is so successful that desert locusts appear to produce dry excretory pellets under desiccating conditions. In fact, the osmotic concentration of the fluid they excrete has been shown to be about 1800 mOsm. By removing water from both the urine and feces (both of which pass through the rectum), these insects achieve greater water savings than are possible in mammals where the feces can only be concentrated to a level isosmotic to the blood.

8.5 Water uptake by arthropods from a subsaturated atmosphere

The above description of the function of the excretory system of the desert locust also seems to be a pretty good description of these functions in most terrestrial insects. In all cases examined to date, the Malpighian tubules produce an isosmotic primary urine which can be modified by downstream portions of the tubules or the gut. Extensive osmotic modification of the urine, however, is generally restricted to the rectum or ileum. These are the sites of urine concentration, principally through water resorption.

The critical feature of the rectum is the presence of a highly concentrated solution in the intercellular spaces. The hyperosmotic solution can then be used to extract water from the excreta. Water movement is always driven by the activity gradient of water between the rectal lumen and the intercellular space. Figure 8.8 is a diagrammatic representation of the anatomical arrangements in the recta of the mealworm, *Tenebrio*. Here, the tips of the Malpighian tubules lie very close to the rectal wall. Both the rectum and the Malpighian tubules are surrounded by an osmotically tight cryptonephridial sheath. Ions transported into the Malpighian tubules through windows in this sheath produce a very concentrated solution that draws water from the rectum. As the insects move air in and out of the rectum, water vapor moves down its activity gradient from air to tubule lumen. This fluid then flows down the tubules to an osmotically permeable region bathed in hemolymph. There, the fluid comes into osmotic equilibrium with the hemolymph. The net effect is that water removed from subsaturated air is moved into the hemolymph and thus into the animal's tissues.

A number of insects can take up water from subsaturated air. This is enormously important for their survival in a terrestrial environment. Although

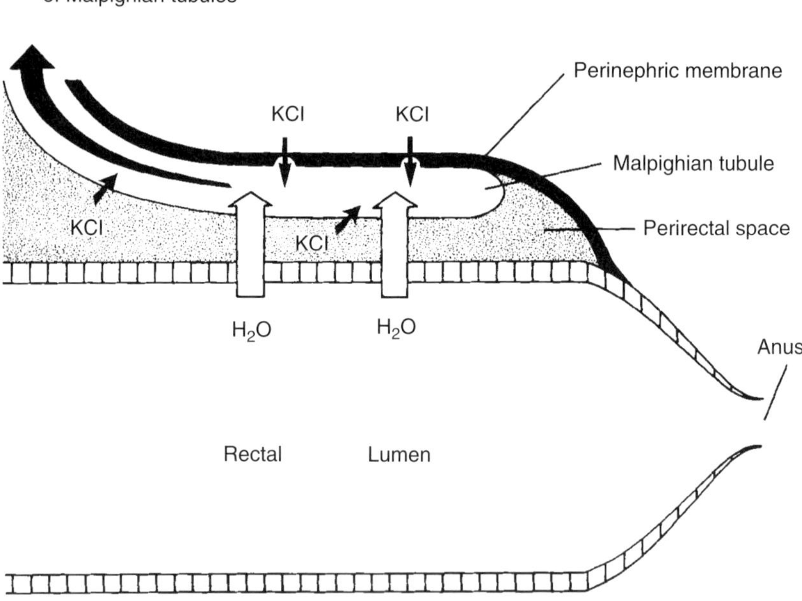

Fig. 8.8. A diagrammatic representation of the anatomical arrangements in the rectum of the mealworm, *Tenebrio*. See text for an explanation on the pathways of water uptake from subsaturated air. (Adapted from Bradley, 1985.)

free-standing water is rare in a number of habitats, short intervals of a high relative humidity are more common. At night, when temperatures drop, the relative humidity rises substantially. In the desert, although falling rain often sinks rapidly into the sand, the humidity of the air is raised by the brief precipitation. It has been shown that insects can take up water from air using the same rectal mechanisms that are used to dry the feces. Instead of taking up water from the urine or feces, these insects can pump external air into and out of the rectum. As long as the activity of water in the air is higher than that in the intercellular spaces in the rectum, water can diffuse down its activity gradient and into the spaces. From there it flows between the cells, ions are removed as described above for the desert locust, and the hemolymph receives a hypo-osmotic fluid.

Table 8.2 shows the threshold relative humidities below which insects of various species are unable to achieve a net uptake of water from subsaturated air. The marked differences in the threshold between species may reflect differences in the types of ions transported into the intercellular spaces, or differences in the rates of water loss from other sites that counter the rate at which water can be obtained from air.

Table 8.2. The order (in boldface) and Genus species (in italics) names of a large number of insects that are able to take up water vapor from subsaturated air

	Critical equilibrium relative humidity (%)
Thysanura (bristletails)	
Ctenolepisma longicauda	60
Thermobia domestica	45
Psocoptera (psocids)	
Liposcelis rufus	58
Orthoptera (grasshoppers, cockroaches)	
Chortophaga viridifasciata	82
Arenivaga investigata	83
Coleoptera (beetles)	
Tenebrio molitor	88
Lasioderma serricorne	43
Siphonaptera (fleas)	
Xenopsylla brasiliensis	50
Xenopsylla cheopis	65
Ceratophyllus gallinae	82

The critical equilibrium relative humidity represents the relative humidity above which the insects can take up water from air. Note that a large number of very diverse insects have the capability of obtaining water in this manner.

A few insects and other arthropods can take up water from subsaturated air at morphological sites other than the rectum. The desert cockroach, *Arenivaga investigata*, produces a hyperosmotic secretion on bladders near the mouth. After this fluid has taken up water vapor from the air, the fluid is swallowed through the mouth. The net effect, of course, is an uptake of water from the air (Table 8.2). Ticks also produce a hyperosmotic secretion from the salivary glands that can be exposed to air and then swallowed. These capacities by arthropods to obtain water from air are of enormous advantage in terrestrial environments where free-standing water can be quite rare.

Suggested additional readings

Bradley, T.J. (1985) The excretory system: structure and physiology. In: *Comprehensive Insect Physiology, Biochemistry and Pharmacology*. G.A. Kerkut & L.I. Gilbert, eds. Pergamon Press, Oxford, NY.

Braun, E.J. & W.H. Dantzler (1997) Vertebrate renal system. In: *Handbook of Physiology*. W.H. Dantzler, ed. Oxford University Press, Oxford.

Garrett, M.A., T.J. Bradley, J.E. Meredith & J.E. Phillips (1988) Ultrastructure of the Malpighian tubules of *Schistocerca gregaria*. *J. Morph.* 195:313–325.

Hadley, N.F. (1975) *Environmental Physiology of Desert Organisms*. Dowden, Hutchinson & Ross, New York.

Hillyard, S.D., K. S. Hoff & C. Propper (1998) The water absorption response: a behavioral assay for physiological processes in terrestrial amphibians. *Physiol. Zool.* 71:127–138.

Hughes, M.R. (2003) Regulation of salt gland, gut, and kidney interactions. *Comp. Biochem. Physiol.* 136:507–524.

Noble-Nesbitt, J. (1998) Hindgut with rectum. In: *Microscopic Anatomy of the Invertebrates*. F.W. Harrison & M. Locke, eds. Wiley-Liss, New York.

Schmidt-Nielsen, K. (1994) *Animal Physiology: Adaptation and Environment*. Cambridge University Press, Cambridge, MA.

Viborg, A.L. & P. Rosenkilde (2004) Water potential receptors in the skin regulate blood perfusion in the ventral pelvic patch of toads. *Physiol. Biochem. Zool.* 77:39–49.

9 Membranes as sites of energy transduction

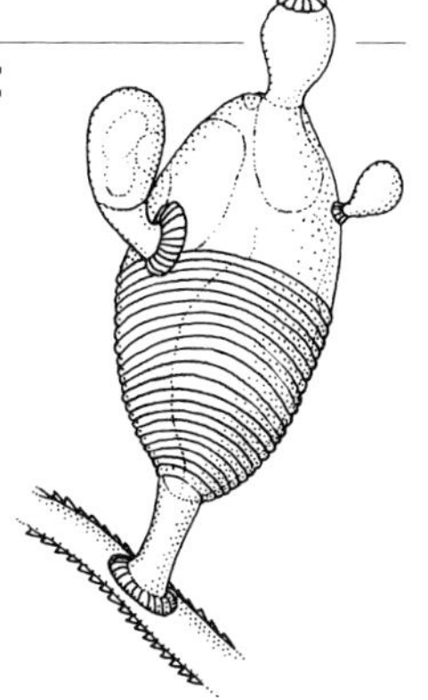

9.1 Introduction

Chapters 1 and 2 discussed the importance of the interactions of solutes with water, and the profound effects these interactions can have on the activity of water. The activity of water and the concentration of specific solutes can in turn influence the structure and function of proteins and membranes. Differences in the activity of water across membranes have important implications for the movement of water because the potential energy contained in the osmotic differences can serve as the driving force for net diffusion across the plane of the membrane.

The types of solute can also be of great biologic significance. In particular, the capacity of charged solutes to separate into ions of opposite charge when dissolved in water is an important characteristic. For example, when table salt (NaCl) is dissolved in water, the majority of the molecules dissociate into positively charged sodium (Na^+) ions and negatively charged chloride ions (Cl^-). The dissolution of the NaCl molecule does not lead to a change in net charge because the number of positive and negative ions is equal, but it does lead to strong local charges around the ions, as discussed in Chapter 2.

The hydrophobic lipid bilayer that forms the central structure of membranes serves not only as a barrier to the diffusion of water, but also as a barrier for the diffusion of solutes, particularly charged solutes such as ions. By restricting the diffusion of ions across the plane of the membrane, the membrane thus serves as an electrical resistor.

Electrical flow is carried by charged particles. In the electrical lines in your house, the electrical flow consists of electrons, each of which carries a charge of −1. Electrical flow can also be carried by other charged particles in the form of ions. In the lead acid battery found in most cars, for example, the electrical energy is stored as hydronium ions (H^+) in the acid and this is transferred to the lead cells creating electrical flow when power is needed from the battery.

In the cells of animals, separation of charge can also lead to electrical potentials and indeed current. Let us imagine two solutions separated by a membrane. Under the initial conditions in our example, each of the solutions has an equal number of positive ions (cations) and negative ions (anions) (Fig. 9.1a). If a few of the cations (in this case sodium ions) are transferred across the membrane without anions following, this leads to a slight difference in the equality of cations and anions on both sides of the membrane and produces a net electrical charge across the membrane. The side to which the sodium ions were transferred develops a net positive charge and the opposite side, where chloride ions predominate, a net negative charge. There is no net flow of electricity, however, because the membrane is acting as a resistor. In the steady-state condition illustrated in Fig. 9.1b, potential energy is stored across the membrane both as a chemical gradient, owing to slightly different concentrations on either side of sodium and chloride ions, and as an electrical charge. This is referred to as an electrochemical gradient because energy is stored in both forms.

Now imagine the same situation, but with the addition of a channel placed in the membrane that allows only sodium to cross the membrane (Fig. 9.1c). Now the potential energy stored across the membrane in the form of an electrochemical gradient drives sodium across the membrane. Chloride ions cannot follow because the channel only allows sodium to pass. The effect will be for a brief period in which sodium ions diffuse across the membrane, driven down their electrochemical gradient. This movement of sodium ions represents an electrical flow, that is, net charges are moving across the membrane. This represents a situation in which chemical potential energy (the separation of sodium and chloride ions) is used to drive ion flow. This is a form of energy transduction in which potential energy in the form of electrical charge is turned into kinetic energy in that the ions are driven in a specific direction.

To the extent, therefore, that membranes can separate chemicals and charged ions on either side of the membrane, these membranes can act as sites of energy transduction. Let us now consider some real-life examples of this energy transduction in action.

9.2 Mitochondrial ATP production

The passage of glucose and other carbohydrates through the pathway referred to as glycolysis results in the production of pyruvate. All of these

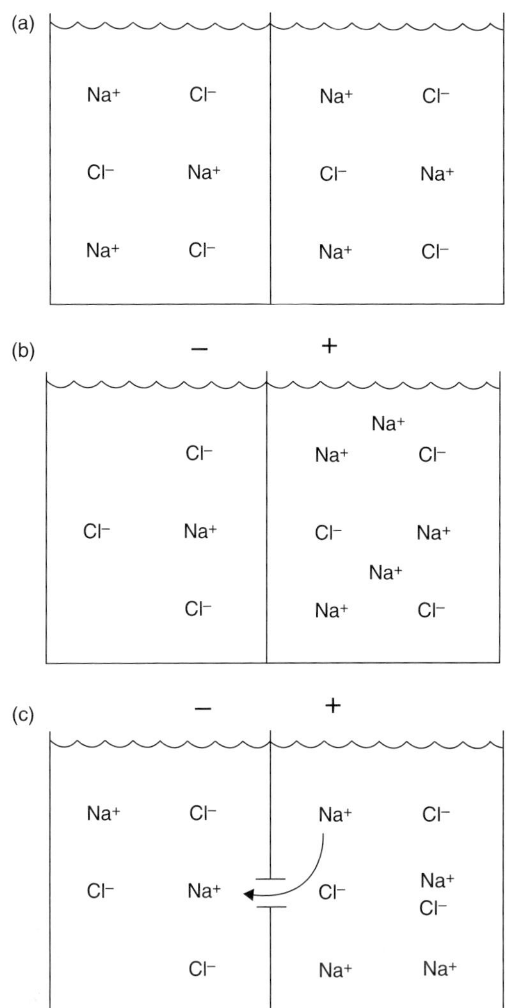

Fig. 9.1. (a) Consider two compartments separated by a membrane. Owing to the even distribution of ions in the two compartments, there is no electrical charge across the membrane. (b) If there is an uneven distribution of ions across the membrane, energy is stored as an electrical potential. (c) The energy stored in the electrical potential can be used to drive ions across the membrane. In this manner, potential energy in the form of an electrical charge can be used to move ions.

reactions occur in the cytoplasm of animal cells. In order for pyruvate to undergo further aerobic metabolism, it must enter the mitochondrion. The mitochondria in eukaryotic cells are surrounded by two membranes (Fig. 9.2). The outer membrane is quite permeable to most compounds and can be ignored for our purpose. The inner membrane is the site of active ion transport and energy transduction (Fig. 9.3) and will be the subject of our discussion here. Pyruvate enters the mitochondrion through specific channels in the inner membrane, combines with malate to form citrate, and cycles through the citric acid cycle. Each passage through the cycle

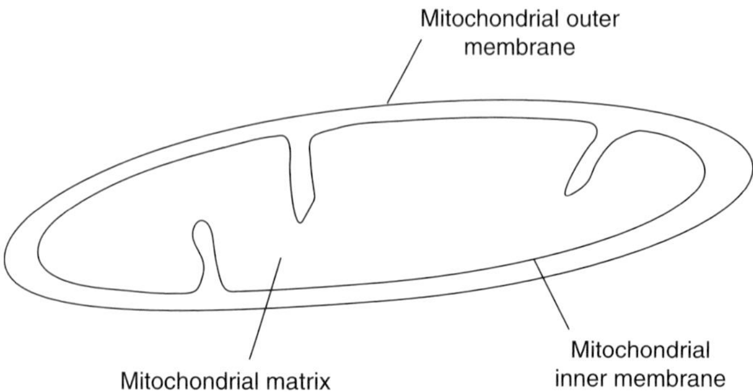

Fig. 9.2. A diagrammatic representation of a mitochondrion. Mitochondria have an inner membrane and outer membrane surrounding the mitochondrial matrix.

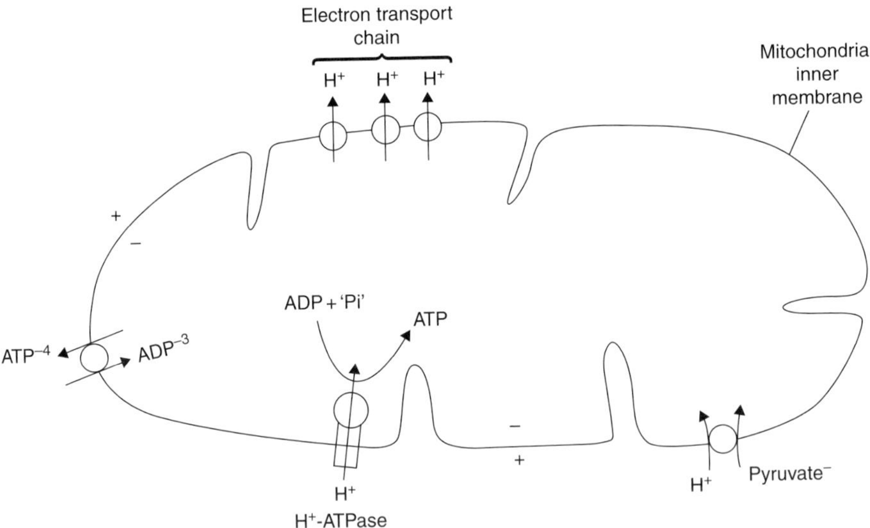

Fig. 9.3. The electron transport chain acts as a pump for hydronium ions, producing electrical and pH gradients across the inner mitochondrial membrane. These gradients are used to produce ATP, drive the exchange of ATP for ADP across the mitochondrial membrane, or import pyruvate into the mitochondrial matrix.

produces three $NADH_2$ and one $FADH_2$. It is these compounds that provide the reducing power for subsequent reactions in the electron transport chain.

The electron transport chain is a series of enzymes and cytochromes that accept electrons from $NADH_2$ or $FADH_2$. These electrons move from cytochrome to cytochrome within the chain. The chain ranges from the upstream end, which is highly reduced by the addition of electrons, to the other end of the chain, which is less reduced (i.e., more oxidized). The electrons can, therefore, flow from the reduced end to the oxidized end of the chain. As they do so, the electrons give up energy because they are flowing from a high-energy state to a more low-energy state. The energy given up by the electrons as they move along the chain is captured by the cytochromes and used to pump hydronium ions (H^+) from inside the mitochondrion (the mitochondrial matrix) to outside the mitochondrion (the cytoplasm). In other words, the electron transport chain consists of three hydronium ion pumps that transport hydronium ions out, and the energy for this transport is provided by electrons moving along the electron transport chain. I have indicated that the end of the chain is less reduced (i.e., more oxidized) and this gradient of reduction is the force that energizes the electron flow. The end of the electron transport chain is maintained in the oxidized state by the actions of cytochrome oxidase. This enzyme combines the use of oxidizing power of oxygen to keep the gradient intact. O_2 accepts the electrons, forming $2O^-$. These combine with $4H^+$ in the cytoplasm to form two water molecules ($2H_2O$). You will remember that oxidative metabolism of carbohydrate produces CO_2 and H_2O.

The chemical gradient of H^+ can also be considered a pH gradient across the membrane because pH is a function of the concentration of H^+ ions. The H^+ ions that are pumped across the membrane by the electron transport chain are unaccompanied by any negatively charged counterion. As a result, there is a large electrical gradient established across the mitochondrial membrane. The combination of the chemical gradient and the electrical gradient is termed the electrochemical gradient. This electrochemical gradient represents a form of potential energy stored across the membrane.

Elsewhere on the mitochondrial inner membrane are molecules referred to as H^+-ATPases. These molecular complexes allow H^+ ions to move back into the mitochondrial matrix through a pore in the inner mitochondrial membrane (Fig. 9.3). The driving force for this movement is the electrochemical gradient for H^+ ions. The H^+-ATPase captures the energy associated with the movement of H^+ ions down their electrochemical gradient to synthesize ATP by means of the reaction

$$ADP + P + \text{energy} \rightarrow ATP \quad (9.1)$$

Energy must be added to the molecules on the left-hand side of Eqn 9.1 to produce ATP because ATP contains more chemical energy in its phosphate bonds than does ADP. In the mitochondrion, that energy comes from the movement of H^+ ion through the H^+-ATPase. You may be aware that the term ATPase is used for molecules that split ATP and use the energy to do work. This could be described using an equation as follows:

$$\text{ATP} \xrightarrow{\text{Ion movement across the membrane}} \text{ADP} + \text{P} + \text{the establishment of an ion gradient}$$

In the above equation, the energy contained in the phosphate bond of ATP is released when ADP+P is formed. Part of the energy is stored in the form of an ion gradient. The H^+-ATPase in the mitochondrion is essentially an ATPase that is running backwards; in other words, it takes the energy in ions moving energetically downhill (negative work) and stores that energy in ATP.

The system I have described above that occurs in mitochondria is an example of energy transduction. The energy stored in the food is directed by a variety of chemical pathways into the production of $NADH_2$ and $FADH_2$ in the mitochondrial matrix. This is not yet a form of energy transduction because energy that was initially chemical potential energy in food is still chemical potential energy. The electron transport chain, however, turns this chemical energy into an electrochemical gradient. This is a form of energy transduction because potential energy stored in chemical bonds is transformed into a electrical gradient. The H^+-ATPase in turn transforms the potential energy stored in the form of the electrochemical back into potential energy stored in the chemical bonds in ATP. In both of these transformation events, a membrane is involved and indeed required for this activity.

The electrochemical gradient of H^+ ion across the mitochondrial membrane is used for other purposes by the mitochondrion as well. For example, the production of ATP is only useful if that ATP can get out of the mitochondrial matrix and into the cytoplasm where the remaining cellular machinery can make use of it. Therefore, there must be a rapid exchange of ADP from the cytoplasm for ATP in the mitochondrion. This is facilitated by a molecular exchange process in the inner mitochondrial membrane that exchanges one ATP molecule for an ADP. ATP has a net charge of –4 while ADP has a net charge of –3. The exchange of one ATP for one ADP, therefore, leads to a net movement of one negative charge out. The electrochemical gradient established by the electron transport chain is the driving force for the exchange of ATP for ADP. Similarly, in order for the mitochondrion to function smoothly and rapidly, pyruvate must constantly enter the mitochondrial matrix. Pyruvate has a net negative charge. Its movement into the mitochondrion occurs through a transport moiety and is driven by cotransport with H+ across the membrane.

We can demonstrate that the electrochemical gradient generated by the electron transport chain really does drive all these processes. If you remove the oxygen from the mitochondrion, then the electron transport chain grinds to a halt because the entire chain becomes evenly reduced. The gradient for H^+ ions decays in the absence of the electron flow and the processes driven by the gradient cease. In another form of evidence, if you introduce a H^+ ionophore into the mitochondrial membrane, you make the membrane leaky to H^+ ions. The gradient disappears and ATP formation ceases. Importantly, the electron transport chain continues because that series of reactions has not been inhibited.

The reactions which occur at the mitochondrial membrane provide an interesting and readily understandable example of energy transduction occurring at a membrane. It is not only pedagogically useful, but also fundamental to an understanding of the biology of membranes. This process for aerobically forming ATP occurs not only in mitochondria, but also in bacteria. It is one of the most fundamental processes for energy storage and maintenance of aerobic life on earth. In addition, the process of photosynthesis in plant chloroplasts is essentially this process in reverse, in which the energy in sunlight is transduced into an electrochemical gradient for H^+ ions, which in turn is used to reduce CO_2 and produce carbon-rich carbohydrates.

We will see in the examples below, and in the next chapter, that energy transduction at membranes is used not only to produce ATP but also in a variety of processes central to osmoregulation in animal cells.

9.3 Vertebrate intestinal epithelial cells

The cells lining the small intestine of vertebrates are termed enterocytes. These cells are the site of uptake of nutrients from the lumen of the intestine following the digestive processes initiated by enzymes secreted by the pancreas. The cells play a major role not only in absorbing nutrients but also in transporting vital ions into the body and absorbing water that otherwise would remain in the gut lumen. The section below will describe some of the processes involved in ion, nutrient, and water absorption from the vertebrate gut.

The enterocytes lining the small intestine are bathed on their basal surface by the interstitial fluids inside the villi. Owing to the permeability of the capillaries, the interstitial fluids are free to exchange, at least with regard to fluids and small soluble compounds, with the blood (Fig. 9.4).

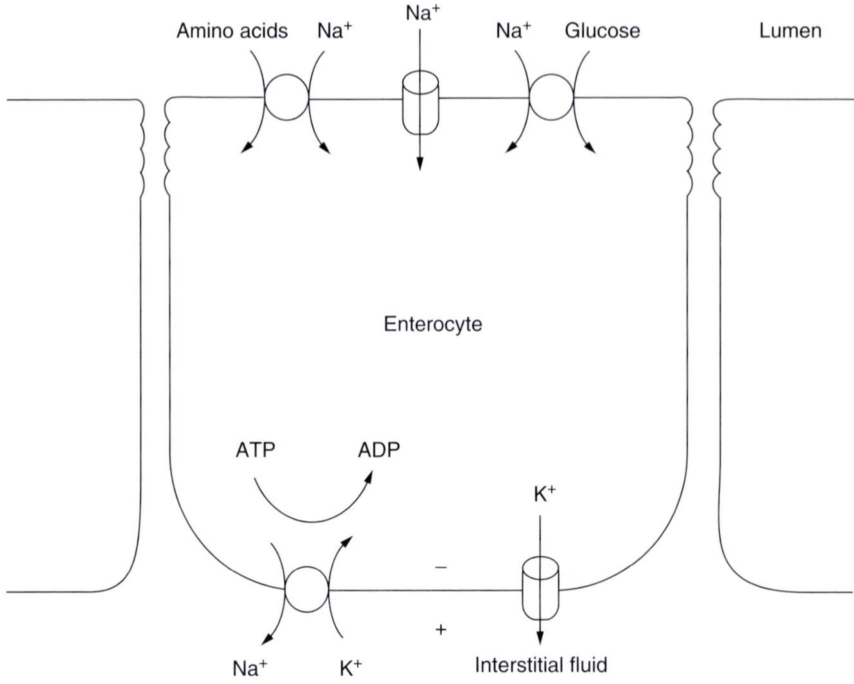

Fig. 9.4. The enterocyte is the site of sodium and nutrient uptake in the intestine of vertebrates. Na^+/K^+-ATPase on the basal membrane removes sodium ions from the cytoplasm and establishes a strong sodium gradient across the apical membrane. This sodium gradient is used to drive the inward transport of sodium ions and also (via cotransport processes) amino acids and glucose.

The enterocytes are bathed on their apical surface by the gut contents. The apical surface is highly convoluted in the form of microvilli which serve to greatly increase the amount of apical membrane in contact with the external gut contents.

The basal membrane of the enterocytes contains the Na^+/K^+-ATPase complex, which is common in animal cells. This enzyme uses the energy contained in the ATP molecule to power the transport of $3Na^+$ ions across the membrane from the cytoplasm to the interstitial fluid. Simultaneously, the enzyme transports $2K^+$ ions from the interstitial fluid to the cytoplasm (Fig. 9.5). The actions of this transport moiety serve to make the sodium concentration in the cell cytoplasm very low, and to enrich the fluid with potassium ions. As the number of charges passing across the membrane in both directions is not equal, the transport also makes the inside of the cell slightly negative relative to the interstitial space. The basal membrane also contains a potassium channel (Fig. 9.4). The activity of the Na^+/K^+-ATPase

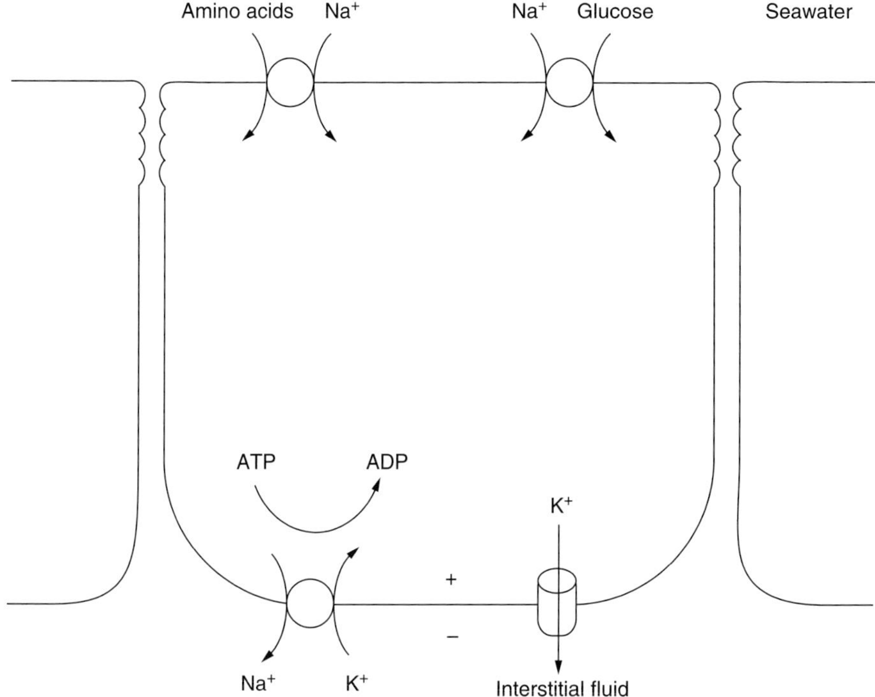

Fig. 9.5. The epithelial cells in the integument of marine larvae also use Na^+/K^+-ATPase to reduce the sodium content of the cell cytoplasm and establish a strong sodium gradient between the cell and the seawater outside. This gradient is used to take up valuable organic compounds from the seawater.

produces a substantial gradient for potassium ions, with inside the cytoplasm having a higher concentration than the interstitial fluid. As a result, potassium can move out of the cell through this potassium channel. As anions cannot follow, the gradient does not go to zero. Instead, enough potassium leaks out to form a substantial electrical gradient (on the order of 60–70 mV inside negative) and the diffusion of potassium slows when the electrical gradient across the basal membrane is strong enough to counter the chemical gradient of potassium ions.

The apical membrane of the enterocyte contains sodium channels (Fig. 9.4). The concentration of sodium in the cytoplasm is very low due to the actions of the Na^+/K^+-ATPase. If the fluid in the gut lumen contains sodium ions, even in relatively low concentrations, these will be able to move down their chemical gradient into the cell cytoplasm. Under many circumstances, the sodium gradient across the apical membrane is actually quite large. If the animal has been drinking the external medium in any waters except strictly freshwater, substantial sodium will be present. If the animal has eaten a meal, the food will be accompanied by saliva, stomach secretions, and secretions from the pancreas. All of these have sodium concentrations equal to the blood, that is, about 100–150 mM in most animals. Therefore, under most circumstances, a substantial sodium gradient exists favoring sodium entry across the apical membrane.

The apical membranes also contain transport moieties that couple the movement of sodium inward across the apical membrane to inward glucose transport (Fig. 9.4). This process is called cotransport. The molecular complex couples the movement of glucose from lumen to cytoplasm to the inward movement of sodium. More remarkably, the energy for this process is the sodium gradient itself. In the absence of the sodium gradient, glucose cannot be transported in this manner. In the presence of the sodium gradient, glucose uptake is stoichiometrically coupled to sodium uptake.

Similar mechanisms present in the apical membranes are involved in the uptake of amino acids (Fig. 9.4). Amino acids can be quite different biochemically from each other. Some are positively charged at physiological pHs, some are negatively charged, and some are neutral. In addition, some such as proline and hydroxyproline have rather large side groups. To date, four distinct classes of amino acid transporters have been isolated from enterocytes. Each one deals with a different class of amino acid. What these transporters have in common is that they all use the sodium gradient to provide the energy for the movement of amino acids across the apical membrane. Once again, in the absence of a sodium gradient, these transport processes cannot occur.

The apical membranes of the enterocytes also contains transport moieties for a variety of ions and micronutrients (e.g., vitamins) that the animal require from its food. Some of these depend on the sodium gradient as an energy source, some use ATP or GTP to power solute uptake. Regardless of the precise mechanism, it is clear that a large variety of solutes are transported from the lumen of the gut into the cytoplasm. The principal one of these is sodium. This ion, once it has entered the cytoplasm, is quickly removed by the Na^+/K^+-ATPase. This maintains the sodium gradient across the apical membrane. Glucose and amino acids accumulate in cell, but as there are channels for these nutrients in the basal membrane (Fig. 9.4) they are able to diffuse down their concentration gradients into the interstitial fluid. Similarly, micronutrients and other ions such as calcium diffuse from the cytoplasm into the interstitial fluid. The net result of this is that a variety of solutes move from the gut lumen, across the cell, and into the interstitial fluid. These processes serve to increase the activity of water in the fluid in the gut lumen which is being depleted of sodium and other ions, sugars, amino acids, and numerous small solutes found in the food. Both the apical and basal membranes of the enterocytes are relatively permeable to water. As a result, water diffuses across the epithelium in proportion to the rate of ion and nutrient transport. As solutes move into the cell water follows. As these nutrients then move into the interstitial fluid, water follows into that space as well. As a result, the retrieval of water from the gut depends on and is driven by the movement of solutes across each of the membranes of the cell.

Let us return to the issue of the energy sources for each of these processes. The energy required to deplete the cell cytoplasm of sodium ions and to enrich it in potassium ions derives from ATP. The transduction of chemical potential energy into an electrochemical gradient is carried out by the Na^+/K^+-ATPase. The basal membrane of the cell also has a large negative electrical potential (inside negative) and this too is powered indirectly by the ATPase. The uptake of sodium and the sodium-coupled cotransport of glucose and amino acids depends on the actions of the Na^+/K^+-ATPase. We can see, therefore, that the vast majority of the transport processes are driven directly or indirectly by the splitting of ATP by Na^+/K^+-ATPase in the basal membrane. Even the transport of molecules against steep gradients at the apical membrane are powered by the splitting of ATP at the basal membrane. Finally, what about water movement? The gut is capable of rapid water uptake from the meal. What is the energy source for that? You are familiar by now with the processes by which water moves across biologic membranes. Water moves due to osmotic forces and it is the movement of solutes that forms the activity gradients. Therefore, the energy for water movement across the gut also derives from the splitting of ATP at the basal membrane by Na^+/K^+-ATPase.

There is one last scenario we should explore before leaving the issue of solute and water movement across the enterocyte. What about the situation in freshwater animals in which they ingest very dilute waters? Similarly, what happens in terrestrial animals when they drink dilute water? Is the process of water entry fundamentally different under these conditions? In the circumstance of water being swallowed which is low in ions and which is not substantially mixed with saliva, it enters the stomach of a vertebrate or the midgut of an invertebrate as dilute, ion-poor fluid. The fluid in the gut will have a high activity of water, while the blood of the animal will have a lower activity due to the normal solutes found in blood. The stomach of vertebrates is somewhat permeable to water and therefore some diffusion of water occurs down its activity gradient. This is a good thing if you are very dehydrated and thirsty because it means that the water rapidly enters your bloodstream. As the water passes further into the intestine, the epithelium is even more permeable than that of the stomach and water rapidly comes into equilibrium with the blood. This means that the water drunk by a vertebrate fairly rapidly moves into the blood stream even if sodium is not present in the gut lumen. A similar situation exists for invertebrates. In most of these animals, no stomach is present so ingested, dilute water rapidly moves into the blood. The process is passive, that is, it does not require simultaneous ion transport and is driven solely by the gradient for the activity of water.

9.4 The integument of marine larvae

The above example of the intestinal cell of vertebrates represents a cell that is highly differentiated. Are similar mechanisms at work in other cell types? Are these mechanisms primitive or merely found in derived cells?

The fundamental, ancient, and indeed universal nature of the energy transduction mechanisms associated with membranes is demonstrated by their presence in a taxonomically diverse array of marine invertebrates. The process has been studied in greatest detail in the larvae of mollusks and echinoderms, although it also occurs in many other marine phyla including annelids and protozoa. Marine larvae are very vulnerable to predation, starvation, and the risk of drifting into adverse environments. There is a premium on rapid growth to a larger and more secure life history stage. In their early developmental stages, the larvae of many marine taxa lack mouths or exist in waters with little particulate food. The larvae obtain organic nutrients and manage to grow under these circumstances by absorbing soluble organic nutrients (dissolved organic carbon) directly from seawater across their integumental cells. Seawater has a surprisingly high content of dissolved organic matter, including amino acids, sugars, and peptide fragments. The larvae of marine organisms often absorb these compounds using the mechanism illustrated in Figure 9.5. The integumental cells of the larvae contain Na^+/K^+-ATPase. This serves to keep the sodium concentration in the cytoplasm very low and to accumulate potassium ions. The apical membranes of the epithelial cells in the integument possess glucose uptake mechanisms that use sodium gradient as the energy source for the cotransport of sodium and glucose. Similarly, amino acids are taken up by mechanisms similar to those described above in enterocytes and these too use the sodium gradient as an energy source. It is clear, therefore, that the role of Na^+/K^+-ATPase in producing an electrochemical gradient for nutrient uptake is extremely widespread and ancient. It was an integral part of the physiological armature of early marine creatures and has been retained as a mechanism linking nutrient uptake to the expenditure of metabolic energy in the form of ATP.

9.5 Conclusions

Chapter 2 emphasized that the movement of water across membranes, as well as into and out of animal bodies, involves the generation of activity gradients for water through the movement of solutes. In this chapter, I have discussed the concept that the movement of ions using ATPases leads not only to the transport of a single solute, but also to the establishment of electrochemical gradients that can serve as sources of energy for a variety of transport processes. By coupling transport activities between the apical and basal membranes, the cells can also create activity gradients for water across an epithelium. The transduction of energy from chemical to electrochemical forms, therefore, provides energy for a variety of vital activities and links these processes across the cell. The following chapter will provide some specific examples of how the appropriate distribution of ATPases, cotransporters, and channels can lead to a surprising array of functions in animals, including ion and nutrient uptake, fluid secretion, regulation of pH, as well as volume and osmotic regulations.

Suggested additional readings

Beyenbach, K.H. (2001) Energizing epithelial transport with the vacuolar H^+-ATPase. *News Physiol. Sci.* 16:145–151.

Faxen, K., G. Gilderson, P. Adelroth & P. Brzezinski (2005) A mechanistic principle of proton pumping by cytochrome c oxidase. *Nature* 437:286–289.

Harvey, W.R. & C.L. Slayman (1994) Coupling as a way of life. *J. Exp. Biol.* 196:1–5.

Peres, A., S. Giovannardi, E. Bossi & R. Fesce (2004) Electrophysiological insights into the mechanism of ion-coupled cotransporters. *News Physiol. Sci.* 19:80–84.

Wieczorek, G.G. & W.R. Harvey (2001) Structure-function relationships of a-, F- and V-ATPases. *J. Exp. Biol.* 204:2597–2605.

Wright, S.H. & D.T. Manahan (1989) Integumental nutrient uptake by aquatic organisms. *Am. J. Physiol.* 51:585–600.

10 Transport of ions and water in epithelia: molecular insights

10.1 Introduction

As indicated in Chapter 9, membranes serve as platforms within which ion transport moieties can reside and function. In some cases these transport moieties are fairly complex, consisting of multiple proteins that must interact in precise ways to achieve active ion transport. The ion transport serves to store energy as ionic and electrical gradients, including the energy gradients required for the movement of water.

In this chapter, I will provide examples of three epithelia whose primary function is osmotic homeostasis and/or water movement. As only three examples will be provided, it is clear that this list is not intended to be exhaustive. Instead, my goal is to provide examples of the breadth and diversity one encounters in animal cells. In addition, I seek to convey the concept that the types of ion-transporting mechanisms available through evolutionary modification, gene replication, and subsequent specialization are limited. However, through the placing of a limited number of types of moieties in new locations, the number and diversity of physiological functions and specialized cells that have been achieved are bewilderingly large.

10.2 Insect Malpighian tubules

Insect Malpighian tubules are the site of urine formation in insects. The tubules are evaginations of the gut. On their basal surface, the tubules are bathed by the insect's blood (termed hemolymph) while the apical surface faces the urinary space. In insects, the urine is formed by active transport of ions from the hemolymph to the urinary space, with water following.

The urine, once formed, flows along the lumen of the Malpighian tubules, into the gut, and eventually is excreted through the anus. In all insects that we are aware of, the primary urine is isosmotic to the blood. Freshwater insects produce a dilute urine by resorbing ions from this primary urine, while terrestrial insects produce a hyperosmotic excreta by resorbing water from the urine.

Rhodnius prolixus, a bloodsucking insect found in North America and South America, can take bloodmeal that is 10 times more massive than its unfed body mass. This meal, being blood, is largely water and the insect excretes about one-half of the volume of the meal in the next hour following the meal. This means that the insect excretes about five times its previous body volume in 1 hr through four Malpighian tubules. This capacity makes the Malpighian tubules of *Rhodnius* the fastest fluid-transporting epithelia known to biology when expressed on a per gram basis.

The cell type that I have chosen to illustrate in detail is the cells of the upper tubule (Fig. 10.1). This is the predominant site of fluid transport and urine formation in the *Rhodnius* Malpighian tubule. This cell type transports fluid very rapidly from the basal fluid (hemolymph) to the apical fluid (urine). To do this, the cell must transport osmotically active ions from the hemolymph to the urinary space. This ion transport reduces the activity of water in the urine, driving water movement across the epithelium.

The apical membrane in the cells of the upper tubule contains a hydrogen ATPase. The ATPase uses the chemical energy in the reaction

$$\text{ATP} \rightarrow \text{ADP} + \text{inorganic P}$$

and couples it to the movement of hydrogen ions from the cell cytoplasm to the urinary space. This transport of H^+ causes the lumen to become more acidic and, because hydrogen ions are positively charged, causes the urinary space to have a positive electrical charge relative to the cytoplasm. The pump, therefore, produces both a chemical gradient and an electrical gradient. These gradients provide an energy source that the cell can use for transporting other ions. In the case of the upper tubule, an ion exchanger is present in the apical membrane of the cell which exchanges a hydrogen ion moving inward down its electrochemical gradient for a sodium and/or potassium ion moving in the opposite direction from the cytoplasm to the urine. This exchanger is termed the H^+/cation exchanger because it can facilitate the outward movement of either Na^+ or K^+. At the same time, chloride ions can move through an apical ion channel from the cytoplasm to the urine, by using electrical gradient as the driving force. The result is that the urine becomes enriched in sodium, potassium, and chloride ions. The driving force for the movement of each of these three ions is the release

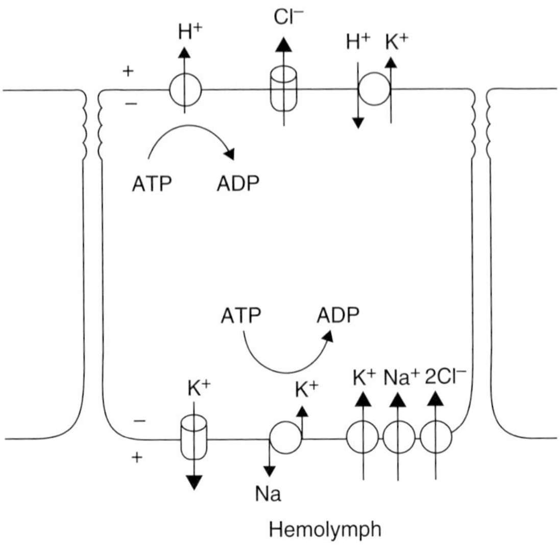

Fig. 10.1. Ion transport activities associated with cells in the upper tubule cells of the Malpighian tubules of *Rhodnius prolixus*. The basal membrane (lower portion of the figure) is energized by a Na^+/K^+-ATPase. K^+ leaks out of the cell through a membrane channel, causing the cytoplasm to have a negative potential relative to the hemolymph. The electrochemical gradient for sodium drives the cotransport of Na^+, K^+, and $2Cl^-$ by an electroneutral cotransporter. Following the subsequent removal of Na^+ by the basal ATPase, the net effect at the basal membrane is the entry of 2KCl. The apical membrane (upper portion of the figure) is energized by a H^+-ATPase. The resulting electrochemical gradient for H^+ can be used to drive the movement of Cl^- through a membrane channel and K^+ through a hydrogen ion exchanger into the lumen. The net movement of KCl across the apical membrane provides the osmotic gradient for water to follow, producing primary urine in the Malpighian tubules.

of chemical energy from ATP, with the intermediate step being the active transport of hydrogen into the urinary space.

There are additional physiological events occurring at the basal membrane of the upper tubule cell. At this location, there is also an ATPase, namely, the Na^+/K^+-ATPase. This transporter uses the chemical energy released by cleaving the phosphate bonds in ATP to transport sodium ions out of the cell against their electrochemical gradient, and potassium ions inward against their electrochemical gradient. A potassium channel in the basal membrane allows these potassium ions to diffuse back out to the interstitial fluid. As the potassium ions are positively charged, their outward diffusion unaccompanied by a negative counterion causes the cytoplasm to have a negative electrical potential relative to the interstitial fluid. Another type of channel in the basal membrane serves to bring ions into the cell. This transporter is referred to as the bumetanide-sensitive $Na^+/K^+/2Cl^-$

cotransporter. The name of this transporter derives from bumetanide, the specific inhibitor that allows one to identify and characterize this transporter, and the types and stoichiometry of the ions that are cotransported, namely, one sodium ion, one potassium ion, and two chloride ions. You will note that the cotransport of these four ions is electroneutral. The ions move across the membrane together, driven by the electrochemical energy contained in the sodium gradient.

The ions delivered by the bumetanide-sensitive $Na^+/K^+/2Cl^-$ cotransporter have different fates within the cytoplasm. The sodium ions can move into the apical urinary space through the H^+/cation exchanger on the apical membrane, or they may be returned to the interstitial fluid by the Na^+/K^+-ATPase on the basal membrane. The potassium ions can move into the apical urinary space through the H^+/cation exchanger on the apical membrane, or they may diffuse back to the interstitial space through the potassium channel in the basal membrane. The chloride ions diffuse into the urinary fluid by means of the chloride channel in the apical membrane.

Both the apical and basal membranes of the Malpighian tubule cells are highly osmotically permeable. The net movement of sodium, chloride, and potassium ions from the cytoplasm to the urinary space reduces the activity of water in the urine and produces a driving force for water to move into the urine. The result of this is the formation of a urine rich in Na^+, K^+, and Cl^-, and isosmotic to the hemolymph.

Let us review where the energy comes from for this rapid fluid transport. At the apical membrane, the energy required for the movement of Na^+, K^+, and Cl^- across the membrane is derived from the electrochemical gradient produced by H^+ transport. This in turn depends on ATP cleavage by the ATPase. At the basal membrane, the Na^+/K^+-ATPase produces the sodium gradient as well as providing the potassium ions, which upon diffusing back out, produce the inside-negative electrical potential. These chemical and electrical gradients combine to provide a strong electrochemical gradient for sodium. It is this gradient that supplies the energy for the inward transport of the ions required for urine formation. In this manner, the two ATPases use chemical energy to produce electrochemical gradients that are used by ion exchangers and cotransporters to move ions across the two membranes. These ion movements, in turn, drive the water movement.

10.3 The mitochondria-rich cells of freshwater animals

As pointed out in Chapter 6, the skin of frogs and toads and the gills of freshwater crustaceans can actively take up sodium and chloride from the external medium. In all of these organisms, the cell type responsible for active uptake of sodium and chloride is termed the mitochondria-rich cell (MR cell). These cells are bathed on their basal surface by an interstitial fluid which is in equilibrium with an extensive blood flow underlying the skin in amphibians or located in the gills in crustaceans. As a result, the fluid at the basal surface of the epithelium has an osmotic concentration around 300 mOsm. The apical surface is bathed by the water passing over the surface of the gills. The osmotic concentration of this fluid can vary, of course, but it is often quite low in both sodium and chloride ions. It is the role of the MR cells to transport sodium and chloride ions against the electrochemical gradient of both ions. In this process, both ionic and osmotic homeostasis are maintained.

The apical membranes of the MR cells contain a H^+-ATPase that serves to produce a strong electrical gradient (Fig. 10.2). This gradient is used to drive

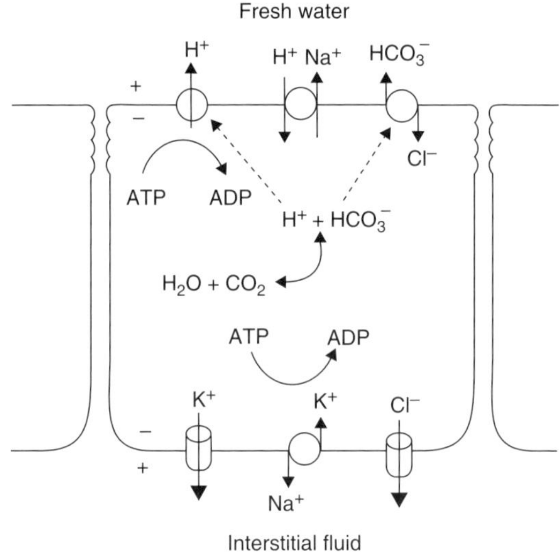

Fig. 10.2. Ion transport activities in the mitochondria-rich cells of freshwater animals. The apical membrane (upper portion of the figure) is energized by a H^+-ATPase. The resulting electrochemical gradient for H^+ can be used to drive the movement of Na^+ through a hydrogen ion exchanger into the cell. Cl^- enters the cell through a Cl^-/HCO_3^- exchanger driven by the negative electrical potential produced by the ATPase in the apical membrane. H^+ and HCO_3^- are produced in the cytoplasm through the chemical interaction of H_2O and CO_2. The basal membrane (lower portion of the figure) is energized by a Na^+/K^+-ATPase. K^+ leaks out of the cell through a membrane channel, causing the cytoplasm to have a negative potential relative to the blood. This electrical gradient drives the exit of Cl^- through a channel from the cytoplasm into the interstitial space. The net effect of the activities at both membranes is the movement of NaCl from the external medium into the animal.

sodium ions inward through a sodium channel in the apical membrane. Chloride ions can also diffuse across the apical membrane and into the cell, but this occurs through an electroneutral mechanism that exchanges Cl^- for bicarbonate (HCO_3^-). The H^+-ATPase tends to acidify the outside aqueous medium. This acidification is neutralized, however, by the export of bicarbonate which results from the Cl^-/HCO_3^-) exchange. The H^+ and HCO_3^- combine in the external fluid to produce water and CO_2.

On the basal membrane, the enzyme Na^+/K^+-ATPase serves to maintain strong gradients across the membrane for both sodium and potassium ions. This process not only serves the usual purpose of charging the membrane, but also serves to move the sodium ions that have entered through the apical membrane on to the interstitial space. Chloride ions are moved from the cytoplasm to the interstitial fluid through a passive chloride channel, which is energized by the electrical potential across the basal membrane. This potential is maintained in these cells, as in most cells, by the outward diffusion of potassium through the potassium channel.

The cells in amphibian skin and crustacean gills, therefore, use ATPases to energize the apical and basal membranes, via chemical and electrical gradients. These gradients in turn provide the energy to actively transport sodium across the apical membrane and chloride across the basal membrane.

The Malpighian tubule cells transport ions rapidly outward, while the amphibian and crustacean cells transport ions rapidly inward. Despite these differences, the mechanisms for charging the membrane and creating electrochemical gradients are similar. The differences arise in the specific positioning of the pumps, ion channels, and exchange processes. These differences may occur at the level of the cell types as illustrated in Figs. 10.2 and 10.3, or they may occur in the form of distribution between the apical and basal membranes. In cells of the Malpighian tubule the bumetanide-sensitive $Na^+/K^+/2Cl^-$ transporter is located in the basal membrane, while in the gills of some marine fish it is located in the apical membrane and is used for sodium uptake from the external medium. The transport moieties that have been found in animal epithelia are very similar across cell types and across animal groups. The various combinations that can be produced by arranging these moieties in different cell types, different membranes, and in different combinations lead to a very large number of combinations and permutations. The physiological differences are produced by placing the pumps and channels in specific locations. The myriad physiological functions carried out by the epithelia of animals (e.g., fluid transport, ions uptake or excretion, pH regulation) are the result.

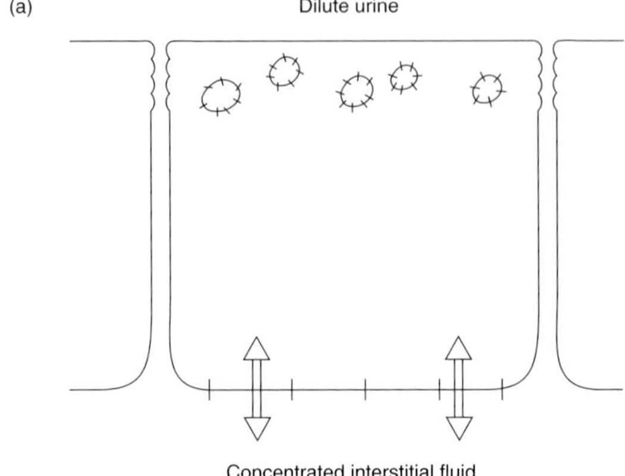

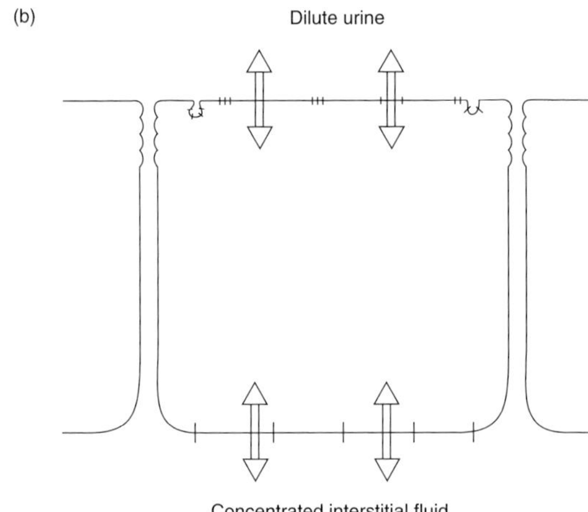

Fig. 10.3. Control of the osmotic permeability of the epithelium in the collecting ducts of the mammalian kidney. (a) The figure illustrates the conditions in the collecting duct when circulating levels of antidiuretic hormone are low in the body. The cells of the collecting duct are shown in their position between the urine and the highly concentrated interstitial fluid in the medulla of the kidney. Aquaporin (AQ) molecules are illustrated as lines in the membrane. The basal membrane contains AQ3 and AQ4 molecules. These molecules greatly increase the osmotic permeability of the basal membrane, causing the cell to always be isosmotic to the interstitial fluid. AQ2 molecules are contained in apical vesicles inside the cell. The apical membrane contains few if any aquaporin molecules, allowing this membrane to have a very low osmotic permeability. As a result, dilute urine flows through the collecting ducts in the kidney and on to the bladder to be excreted. (b) The figure illustrates the conditions in the collecting duct when circulating levels of antidiuretic hormone are high in the body. Under these conditions, the vesicles that contain the AQ2 molecules have been transported to the apex of the cell where they have fused with the apical membrane. As a result, the AQ2 molecules are now inserted in the apical membrane. This markedly increases the osmotic permeability of the apical membrane and, as a result, the whole epithelium. Water can now move down its activity gradient from the urine to the interstitial fluid. In this manner, water is resorbed from the urine and retained in the body, and the urine is made highly concentrated.

10.4 The collecting tubule of the mammalian kidney

The example of the Malpighian tubule cell given above involves a highly permeable epithelium. In such epithelia, where fluid transport is isosmotic, the rate of ion transport directly and linearly determines the rate of fluid transport. The cells in amphibian skins and crustacean gills are, in contrast, relatively impermeable. As a result, ion transport occurs with a disproportionately small quantity of water and any transported fluid is hyperosmotic to the interstitial fluid.

There are additional epithelia in which the appropriate function in maintaining homeostasis involves the capacity to have not a fixed osmotic permeability, but rather a variable one. The example I have chosen for this type of epithelium is the epithelium in the collecting tubule of the mammalian kidney. As described in Chapter 8, the cells in the loop of Henle produce a highly concentrated interstitial fluid in the medulla of the kidney of many mammalian species. As shown in Fig. 10.3a, the concentration of the interstitial fluid in the medulla of the kidney can reach concentrations in excess of 1000 mOsm or more. The cells in the medulla of the kidney, and this includes epithelial cells in the loop of Henle, collecting tubule, and capillaries, are all isosmotic to the interstitial fluid. They survive under these conditions by producing compatible solutes such as proline, leucine, and valine.

The apical membranes of the collecting tubules can have a very low permeability to water. This is achieved using the lipid bilayer which by itself has a very low permeability to water. Apparently, the proteins in the apical membrane are also either present in very low numbers or fairly impermeable to water.

The collecting tubules are capable of changing their overall osmotic permeability. This process is controlled by antidiuretic hormone (ADH), also called as vasopressin. Details of the homeostatic mechanisms controlling the concentration of ADH in the blood are discussed in the next chapter. In this discussion, we will concentrate on events in the cells of the collecting tubule when the circulating levels of ADH increase.

Under conditions of low-ADH concentration in the blood, the apical cytoplasm of the cells of the collecting tubules contain vesicles that have a type of aquaporin molecule (AQ2) inserted into their bilayer. These vesicles remain "parked" in this position as long as ADH levels in the blood are low. If ADH levels in the blood should rise, however, ADH receptors on the basal membrane would experience an increased frequency of occupation by ADH. This increase in occupancy leads to the activation of G proteins in the membrane which diffuse in the lipid bilayer to adjacent enzyme complexes. These complexes, referred to as adenylate cyclase, catalyze the formation of cyclic AMP (cAMP) from ATP. Elevated levels of cAMP cause a cascade of enzymatic reactions leading to the phosphorylation of proteins, and the activation of cytoskeletal elements in the cells. The outcome of this cascade of activation is that the cytoskeletal elements transport the membrane vacuoles that are parked in the apical region of the cytoplasm to the apical membrane, facilitating the fusion of these vacuolar membranes with the apical membrane. That is, the vacuolar membranes containing the aquaporin molecules actually become part of the apical membrane. Upon insertion of the aquaporin molecules, the apical membrane becomes much more permeable to water. As the apical membrane in the unstimulated cell is the major barrier to transepithelial water movements, the insertion of aquaporin into the apical membrane increases the osmotic permeability of the epithelium as a whole. As a result, the large osmotic gradient that exists across the epithelium now can drive the movement of water. Using the aquaporin molecules, water moves down its activity gradient from the urine to the interstitial spaces. In this manner, the level of ADH in the blood controls the permeability of the collecting tubule epithelium, through the action of aquaporin insertion.

10.5 On the more general distribution of aquaporins

As pointed out in Chapter 3, the lipid bilayer in biologic membranes is actually rather impermeable to water. The presence of proteins in the membranes increases the permeability of the membranes to water but only in rather nonspecific and haphazard ways. Aquaporins are protein complexes that greatly increase the permeability of membranes to water, but with little effect on the permeability with regard to solutes. Some cell types, particularly in the integument of animals, carry out their physiological function by being as osmotically impermeable as possible. Others, particularly those involved in transporting fluids, can best carry out their function by being highly permeable to water. It has become clear that almost all of these membranes contain aquaporin molecules that serve to increase water permeability, thereby facilitating the rapid movement of water.

Let us revisit the example of the collecting tubule to illustrate this point. The capacity of the cells of the collecting tubule to insert aquaporin molecules controls the rate of water movement because it converts an impermeable membrane to a more permeable one. The basal membrane is always highly osmotically permeable so it provides no significant barrier to water movement. Why is that basal membrane so permeable? It has been shown that the basal membranes also contain aquaporins. These aquaporins are of different types (AQ3 and AQ4) and are not inserted and removed in response to ADH. To date, aquaporins have been found in a large variety of cells from almost all animal and plant groups. Aquaporins serve to increase the osmotic permeability of bilayers. In all cells, the number and distribution of aquaporin molecules is presumably under genetic control, and in some cell types (such as the collecting tubule) the number of aquaporin molecules in specific membranes can be rapidly modified using membrane fusion and/or internalization.

Suggested additional readings

Hiroi, J., S.D. McCormick, R. Ohtroni-Kaneko & T. Kaneko (2005) Functional classification of mitochondria-rich cells in euryhaline Mosambique tilapia (*Oreochromis mossambicus*) embryos by means of triple immunofluorescence stain for Na^+/K^+-ATPase, $Na^+/K^+/2Cl^-$ cotransporter and CFTR anion channel. *J. Exp. Biol.* 208:2023–2036.

Ianowski, J.P., R.J. Christenson & M.J. O'Donnell (2004) Na^+ competes with K^+ in bumetanide-sensitive transport by Malpighian tubules of *Rhodnius prolixus*. *J. Exp. Biol.* 207:3707–3716.

Kirschner, L.B. (2004) The mechanism of sodium chloride uptake in hyperregulating aquatic animals. *J. Exp. Biol.* 207:1439–1452.

Knepper, M.A., S. Nielsen & C-L. Chou (2007) Physiological roles of aquaporins in the kidney. *Curr. Top. Membr.* 51:122–155.

11 Volume and osmotic regulation

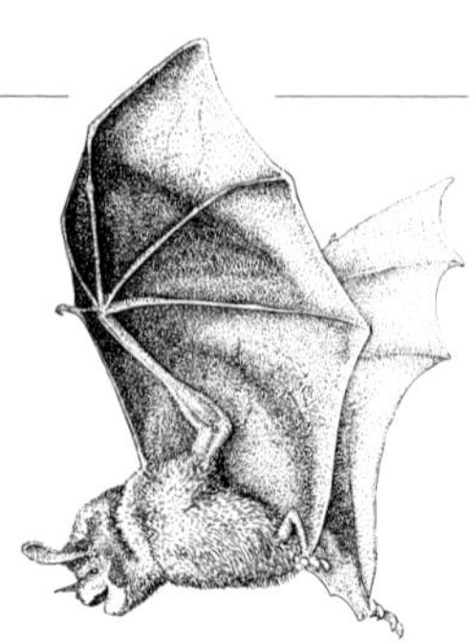

11.1 Introduction

In the preceding chapters, we have examined the capacity of animals to regulate the activity of water inside their bodies in a manner that preserves the structural and functional integrity of their membranes and proteins. We also explored the conflicts that organisms face from the duel challenges of regulating internal osmolality and dealing with potential changes in cell volume.

The first step in maintaining homeostasis involves sensing any changes in the environment and in the internal physiological state. How can an animal measure the activity of water in its internal fluids, and/or the volume of its cells? The sensing and indeed quantification of change is a vital first step in maintaining homeostasis. The second step is to initiate the physiological processes that regulate the perturbed parameter. The third step is to regulate these mechanisms at a strength and duration, which brings back the organism into a range that is physiologically appropriate.

The mechanisms by which animals sense the osmotic concentration of their internal fluids are complex and vary in phylogenetically diverse groups. Over evolutionary time, the sensors, mechanisms, and regulatory processes in animals have become diverse and complex. Physiologists are still in the early stages of exploring regulation, control, and feedback in osmotic control systems. Nonetheless, our current knowledge does give us insights into the processes by which animals can sense and regulate the osmotic concentration of body fluids.

11.2 Cell volume regulation

11.2.1 Responses to cell swelling

Most of the research on cell volume regulation has employed mammalian cell types, including isolated liver cells, red blood cells, and fibroblasts. However, studies with cells from other species tend to demonstrate that the processes found in mammalian cells are ancient and occur in animal cells from all phyla. I will review here some of the processes that have been elucidated that permit cells to regulate their volume in the face of rapid changes in external osmotic concentration.

Let us begin with the example of a cell that finds itself in an extracellular fluid in which the osmotic concentration is falling (Fig. 11.1). Under these circumstances, the cell would begin to swell due to the osmotically driven entry of water. This increase in volume begins to stretch the cell membrane. In many cell types, there are stretch-activated potassium channels located in the cell membrane that open when the cell swells. As the concentration of potassium is much higher inside the cell than in the extracellular fluid, these channels permit potassium ions to flow out, thereby lowering the solute concentration in the cell. As potassium diffuses out, the gradient for water entry is reduced, leading eventually to osmotic equilibrium. The gradient for potassium across the cell membranes remains large, however. The continued diffusion of potassium ions out of the cell now causes the cell to begin to shrink, returning it to its original volume. At that point the stretch-activated channels close and the cell has achieved a return to its normal volume.

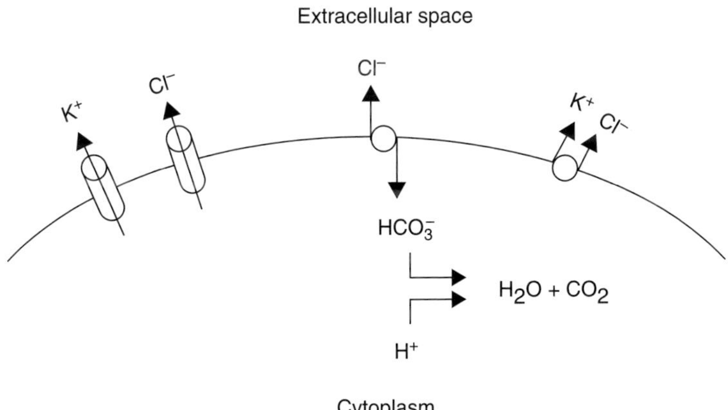

Fig. 11.1. Volume regulatory responses in cells subjected to swelling. See text for further explanation.

Many cells also possess swelling-activated chloride channels. If the intracellular chloride activity exceeds that in the extracellular fluid, this leads to the passive release of additional solutes from the cell. If chloride efflux exceeds potassium efflux, the cell membrane will be depolarized, leading to the opening of voltage-sensitive potassium channels. Finally, some cells, including those of lower vertebrates, contain channels that cotransport potassium and chloride together in an electroneutral process. The net effect of all these events is the rapid and controlled release of both potassium and chloride, the major free solutes in the cell, in response to cell swelling.

In many cells, organic solutes constitute a major portion of the overall solute concentration. The cells of marine osmoconformers possess amino acid channels that are activated during cell swelling. Amino acids released in this manner include taurine, proline, and alanine. The release of amino acids into the blood reduces the number of cell solutes and, thus, cell volume. Mammalian renal inner medullary cells have been found to possess channels for the release of compatible solutes, betaine and sorbitol. This allows the cells to regulate cell volume with little effect on metabolic function.

The activation of the channels promoting solute release need not be a direct physical response to the stretching of the membrane or to membrane depolarization. Some channels are opened in response to intracellular signals such as calcium concentration or cytoplasmic pH. Membrane stretching can also activate G proteins and phospholipases that, in turn, activate channels or promote the insertion of channels into the membranes. In recent years, a great deal of attention has been given to the possible role of the cytoskeleton, particularly actin filaments, in regulating cell volume. The cytoskeleton plays a central role in determining cell shape. Through its attachment to membrane proteins, the cytoskeleton is also well positioned to sense changes in cell volume and stress on the membrane. It has been proposed that changes in tension in the cytoskeleton generated during cell swelling might allow the cytoskeleton itself to play a role in controlling the activation of ion channels and thus the regulation of the volume response.

Clearly, cellular responses to volume changes can be both structurally and biochemically complex. The common factor in all cell types and across phylogenies is the need of the cells to rapidly release solutes in response to rapid increases in cell volume. This is carried out via solute channels that permit the passive efflux of osmotically active solutes from the cell cytoplasm.

11.2.2 Responses to cell shrinkage

Cells must equally well be able to respond to externally increasing osmotic concentrations. In this circumstance, the cells lose water and shrink (Fig. 11.2). This can lead to impaired function in cells such as muscles that

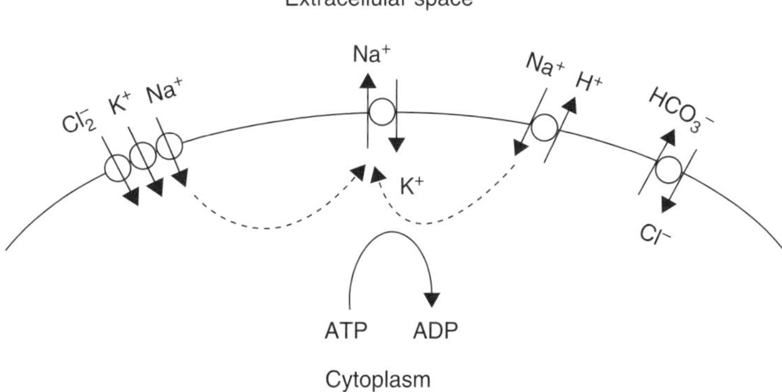

Fig. 11.2. Volume regulatory responses in cells subjected to shrinkage. See text for further explanation.

employ intracellular molecular movements, or nerves that must transport compounds and organelles through the cytoplasm. In addition, loss of cell water increases the concentration of all of the internal solutes. This can change the concentration of small regulatory compounds such as calcium and ATP. It also affects the thermodynamic activity of proteins if their ability to change conformation is affected.

Many cells respond to cell shrinkage by opening the bumetanide-sensitive $Na^+/K^+/2Cl^-$ channel in the cell membrane (see Chapter 10). This channel uses the sodium gradient to energize the simultaneous movement of one sodium ion, one potassium ion, and two chloride ions into the cell. This serves to increase the internal solute concentration, thereby promoting the inflow of water, and restoring cell volume. Ultimately, the energy for the entry of these ions is derived from the Na^+/K^+-ATPase, which maintains the sodium gradient. As this ATPase removes the sodium and replaces it with potassium, the net effect of the activity of the $Na^+/K^+/2Cl^-$ channel and the Na^+/K^+-ATPase is the entry of potassium chloride into the cell cytoplasm. When cell volume has returned to normal, the permeability of the bumetanide-sensitive channel is reduced. A second channel, referred to as NHE, is found in many membranes. It can be activated by cell shrinkage. This channel exchanges hydronium ions for sodium. In response to cell shrinkage, this channel allows sodium to enter, exchanging this ion for H^+ in an electroneutral manner. The N^+/K^+-ATPase exchanges this sodium for potassium. In most cells, the channels that respond to cell shrinkage are regulated by enzymes. The channels frequently are activated by kinases and inhibited by phosphorylases. In some systems, the channels have been shown to be inserted when their activity is needed.

11.3 Sensing osmotic concentration

Now that we have a better understanding of how cells respond in a homeostatic manner to changes in osmotic concentration, we are in a better position to understand the manner by which whole organisms can sense changing osmotic conditions.

In many organisms, including marine animals, crustaceans, vertebrates, and many insects, the principal extracellular solute is sodium. Therefore, cells, including nerve cells that can sense changes in sodium concentration, can provide good information about changes in total osmotic concentration. Most cells possess one or more types of sodium channel. The entry of sodium into the cell tends to depolarize the cell membrane. You are actually personally familiar with such cells. The salt-sensitive taste buds on your tongue have such a sodium-sensing capability. Nerves with these capabilities can be used to sense changes in extracellular water activity based on sodium concentration.

Changes in cell volume are also a mechanism by which animals can detect changes in the osmotic concentration of their body fluids. Most organisms possess neurons and/or cells that respond to changes in cell volume or the volume of specific body compartments through stretch-related sensing processes. In vertebrates, stretch receptors monitor blood volume. In insects, stretching of the body wall initiates processes that control diuresis and thus fluid removal. These organs that monitor solute concentration, especially sodium, as well as the volume of fluid compartments are vital to the homeostatic response of animals in changing osmotic environments.

11.4 Osmotic homeostasis in insects

As described in previous chapters, insects produce a primary urine in the Malpighian tubules through the process of active ion transport via an osmotically leaky epithelium. The resulting fluid is usually isosmotic with the hemolymph and enriched in particular ions (e.g., potassium and chloride) that are actively transported. The urine is then modified prior to excretion in the downstream portions of the Malpighian tubules or in the hindgut.

In this section, I will describe our current understanding of the regulatory and control systems for osmoregulation in three insect species. Although many other insect species have been explored, these three examples have been studied in considerable detail and they provide some insights into the diversity of physiological processes existing in the insects as a group.

In Chapter 7, it was pointed out that Desert Locusts (*Schistocerca gregaria*) are extraordinary in their capacity to produce a concentrated urine and survive in dry, inhospitable environments. At the same time, in periods in which the desert blooms and is filled with succulent green growth, the insects can feed voraciously and pass large volumes of dilute fluids through their excretory systems. Control of diuresis in the Malpighian tubules is controlled by hormones in the hemolymph. Sensation of osmotic and ion concentrations in the hemolymph is carried out in the nervous system and hormones are released by neurosecretory cells in the insect's central nervous system.

Several hormones have been shown to stimulate the rate of urine production in the Malpighian tubules. When feeding on well-hydrated green leaves, the insects take in a large amount of relatively dilute material. When this material enters the gut, ions and nutrients are taken up and water follows. Resulting changes in the ionic concentrations in hemolymph as well as hemolymph volume initiate diuresis. Two peptide hormones, referred to as locust diuretic hormone and locustakinin, have been shown to be rapidly released into the hemolymph from the neurosecretory cells following feeding. These hormones can act individually and as well as synergistically in promoting active ion transport in the Malpighian tubules. As this fluid flows down the Malpighian tubules into the gut, it passes into the ileum. As discussed previously, this organ can resorb ions with water following isosmotically. In this process, the fluid provided in the meal is recycled through the Malpighian tubules and ileum back into the hemolymph. In the process, wastes from metabolism and toxins contained in the meal are concentrated in the urine. If a dilute urine is to be formed, ions are resorbed from the urine prior to excretion. If the insects are faced with desiccating conditions, a concentrated urine can be formed as described in Chapter 7. The control of ion transport in the ileum and rectum is controlled by a hormone separate from the one that controls Malpighian tubule function. In this manner, the process of urine production in the Malpighian tubules can be controlled separately from the processes that modify the urine in the hindgut. It has been shown that the hormones controlling these two systems are distinct, and have different response times appropriate to their distinct functions (Coast et al., 1999).

11.4.1 The bloodsucking bug, Rhodnius prolixus

Insects in the species *R. prolixus* feed on the blood of mammals and birds. The insects feed on no other food throughout their lives. When they do feed, they take on a huge liquid meal of blood, often increasing their body weight by 10-fold. Following the meal, the insects must rid themselves of the large amount of fluid contained in the meal in order to concentrate the nutrients in the meal. *Rhodnius* is capable of removing most of the excess fluid contained in the meal within about 2 hr following the bloodmeal. They achieve this using only four Malpighian tubules. As a result, the Malpighian tubules have the highest rate of fluid transport per gram of any tissue studied to date.

Studies of the control of diuresis in *Rhodnius* have revealed two distinct compounds that serve as diuretic hormones, stimulating ion transport across the epithelium of the Malpighian tubules. A peptide hormone or hormones is released from neurosecretory cells in response to abdominal stretching following the huge meal ingested by these insects (Maddrell, 1969; Maddrell et al., 1991). In addition, the amino acid derivative 5-hydroxytryptamine (5-HT) has been shown to circulate in the hemolymph following the bloodmeal and to be a stimulant for Malpighian tubule secretion. 5-HT has also been shown to stimulate resorptive functions in the downstream region of the Malpighian tubules, a function that produces the copious dilute urine needed to restore volume and osmotic homeostasis following ingestion of the bloodmeal. Once the appropriate amount of fluid has been jettisoned from the insect, it is vital that the rapidly secreting Malpighian tubules reduce their rate of transport. It has been shown that cardioacceleratory peptide 2b only has this function in *Rhodnius*, apparently acting through the formation of cyclic GMP in the cells of the Malpighian tubules.

In *Rhodnius*, as in many insects, therefore, distinct hormones are present that either stimulate or retard fluid secretion by the Malpighian tubules. The presence of multiple regulatory compounds affords the insects with flexibility in response as well as providing backup systems for the control of transport.

11.4.2 The tobacco hornworm, Manduca sexta

The Malpighian tubules of the tobacco hornworm respond to circulating levels of several peptide hormones, including octopamine, leukokinin I, tachykinin-related peptides, and two insect neurohormones, cardioacceleratory peptides 1a and 1b. It is surprising that so many compounds have been identified that affect Malpighian tubule function in *Manduca*. There is a tendency in endocrinology to think of specific functions in a single cell type as responding to circulating levels of single hormone. As discussed by Skaer et al. (1999), studies in insects suggest instead that a constellation of hormones in the blood may influence tissue function. Single hormones can influence the function of more than one tissue. An example is provided by the cardioacceleratory peptides that stimulate both heart rate and urine formation in *Manduca*. In addition, a single tissue can respond to multiple hormones that can have antagonistic or synergistic effects. The overall signal deriving from the combination of compounds provides improved subtlety in response, and permits morphologically separated tissues to be coordinated in their actions.

11.5 The mammalian kidney

In the foregoing chapters we have discussed a great variety of animals in which specialized organs exist that are capable of producing a very dilute or very concentrated secretory fluid. These organs are often distinct from those that produced the primary urine. In fish, for example, the urine is produced in the kidney, but osmotic regulation is largely dependent on specialized cells in the gills. In insects, the primary fluid is produced in one organ, the Malpighian tubules, but is then modified in another, the hindgut. In mammals, by contrast, both urine production and osmotic regulation occur in the same organ, the kidney. The details of this function have been discussed in Chapters 7 and 8. Here I will describe the regulatory processes that control the production of urine with appropriate osmotic concentration, thereby serving to maintain osmotic homeostasis.

In Chapter 8, we discussed at length the capacity of the mammalian kidney to produce a very dilute or highly concentrated urine depending on the physiological needs of the animal. Mammals must be capable of varying the osmotic concentration of the urine and therefore of rapidly adjusting transport processes. For example, a well-hydrated animal that begins its journey across a hot, dry landscape will rapidly need to initiate the processes that retain water and excrete ions. If, after a period of time, that animal comes upon a source of water and drinks a large volume, the water taken up across the gut would initially be used to replace lost blood volume, but the excess would need to be excreted in the form of dilute urine. We can see, therefore, that mammals require a means of rapidly assessing slight changes in the osmotic concentration of the body fluids, and then rapidly adjusting the excretory products to achieve osmotic homeostasis.

The first requirement in any such homeostatic system is the capacity to measure changes in osmotic concentration. Critical measurements in this regard are made by neural cells in the hypothalamus, a region at the base of the brain. The cells here are sensitive to changes in osmotic and/or sodium concentration. When the neurosecretory cells sense an increase in the osmotic concentration of the extracellular fluid that exceeds the set point, the cells begin to release antidiuretic hormone (ADH), also known as vasopressin. This hormone circulates in the general blood circulation. Cells in the collecting duct of the kidney possess receptors that bind ADH. When the receptors are sufficiently occupied by the hormone, they activate a G protein which in turn activates adenylate cyclase. This enzyme catalyzes the formation of cyclic AMP from ATP. Elevated levels of cyclic AMP promote the insertion of aquaporin molecules into the apical membrane of the cells lining the collecting tubules. Because the medulla of the kidney has an elevated osmotic concentration (see Chapter 8), water is drawn from the urine and returned to the mammal's body fluids. This results in the production of concentrated urine, which can then be excreted from the body.

The retention of water serves to dilute the body fluids. If the return of water to the body is sufficiently great, the osmotic concentration of the body fluids will be returned to a value within the set points. At this point, the cells in the hypothalamus reduce the release of ADH and the circulating levels of the hormone drop. As the ADH receptors on the collecting tubules experience lower levels of circulating ADH, the activity of adenylate cyclase is reduced and the aquaporin molecules are removed from the apical membrane. As a result, the osmotic permeability of the collecting tubules drops and less water is removed from the urine passing by.

Thus, the ADH system of mammals represents a classic negative feedback system. If the osmotic concentration goes up, hormone is released promoting water retention. As a result, osmotic concentration drops back into the regulated range. Unlike what has been described for insects, this is not a system that involves multiple hormones having either a diuretic or an antidiuretic effect. Mammals also possess hormones that promote ion secretion into the urine (e.g., aldosterone), but these are more directly involved in ionic regulation than in osmotic regulation.

Mammals, similar to all other animals, must deal with volume regulation as well as osmotic regulation. Regulation of blood volume is critical in mammals because cardiac, lung, and kidney functions all depend on adequate blood volume and pressure. The hypothalamus receives input from arterial and atrial baroreceptors, cells that sense stretching in these vessels. If blood pressure becomes too high, these signals can inhibit the release of ADH, thereby reducing the resorption of water. Increased urine production results in a reduction in blood volume and a return to blood volume homeostasis.

11.6 Summary

This chapter is intended to give you a sense of how the various mechanisms employed by animals to maintain osmotic homeostasis can be controlled. The first challenge is to accurately measure both changes in osmotic concentration and changes in the volume of various body fluid compartments. We have discussed several mechanisms by which this task can be completed. The next task is to adjust the output of osmotic regulatory mechanisms in a manner that returns the values to ones within the set point range. In some animals such as insects, this is achieved by a multiplicity of interacting and synergistic hormones, the sum total of which determine urine output and the degree of ion retention. In mammals, the system is less complicated due to the lack of diuretic hormones, but the system still has the capacity for independent regulation of osmotic concentration and fluid compartment volume.

Our understanding of the control of physiological systems is still in its infancy. In particular, our understanding of the complexity and interactive nature of hormonal signals is expanding rapidly. This is particularly true in the area of osmotic regulation.

Suggested additional readings

Chamberlin, M.E. & K. Strange (1989) An isosmotic cell volume regulation: a comparative view. *Am. J. Physiol.* 257:C159–C173.

Coast, G.M., J. Meredith & J.E. Phillips (1999) Target organ specificity of major neuropeptide stimulants in locust excretory systems. *J. Exp. Biol.* 202:3195–3203.

Eigenheer, R.A., S.W. Nicholson, K.M. Schegg, J.J. Hull & D.A. Schooley (2002) Identification of a potent antidiuretic factor acting on beetle Malpighian tubules. *Proc. Natl Acad. Sci. USA* 99:84–89.

Hoffmann, E.K. & J.W. Mills (1999) Membrane events involved in volume regulation. In: *Membrane Permeability*. D.W. Deamer, A. Kleinzeller & D.M. Fambrough, eds. *Current Topics in Membranes* 48:123–196. Academic Press, San Diego, CA.

Maddrell, S.H.P., W.S. Herman, R.L. Monney & J.A. Overaton (1991) 5-Hydroxytryptamine: a second diuretic hormone in *Rhodnius*. *J. Exp. Biol.* 156:557–566.

Quinlan, M.C., N.J. Tublitz & M.J. O'Donnell (1997) Anti-diuresis in the blood-feeding insect *Rhodnius prolixus Stal*: the peptide CAP2b and cyclic GMP inhibit Malpighian tubule fluid secretion. *J. Exp. Biol.* 200:2363–2367.

Shafer, J.A. (2004) Renal water absorption: a physiologic retrospective in a molecular era. *Kidney Intl.* 66:520–527.

Skaer, N.J.V., D.R. Naessel, S.H.P. Maddrell & N.J. Tublitz (2002) Neurochemical fine tuning of a peripheral tissue: peptidergic and aminergic regulation of fluid secretion by Malpighian tubules in the tobacco hawkmoth *M. sexta*. *J. Exp. Biol.* 205:1869–1880.

Index